DATE DUE

ARCTIC OIL

**Canadian Cataloguing in
Publication Data**

Livingston, John A., 1923–
    Arctic Oil

Based on the program Arctic oil,
from the CBC television series
The Nature of things.
Bibliography: p. 147
Includes index.

ISBN 0-88794-092-7

1. Petroleum industry and trade—
Environmental aspects—
Northwest Territories.
2. Petroleum industry and trade—
Environmental aspects—Yukon
Territory. 3. Petroleum industry
and trade—Social aspects—
Northwest Territories.
4. Petroleum industry and trade—
Social aspects—Yukon Territory.
5. Conservation of natural
resources—Northwest
Territories. 6. Conservation of
natural resources—Yukon
Territory. I. The Nature of things
(Television program). II. Title.

HD9574.C23N67
333.8'232'097199   C81-094230-5

Published by CBC Merchandising
for the
Canadian Broadcasting Corporation.

CBC Merchandising
Canadian Broadcasting Corporation
Box 500, Station A
Toronto, Ontario
M5W 1E6

Design: Keith Abraham
Printed in Canada
5 4 3 2 1 81 82 83 84 85

Photo credits

All photos copyright © the
photographer.

James Murray: Jacket front;
p. 29, p. 105; Photos No. 3, 4,
5, 6, 10, 12, 13, 16, 17, 19,
22, 23, 26, 31, 32, 33, 34, 35,
37, 39, 40.
John Livingston: Photos No. 1
& 11.
*Copyright: Tim Fitzharris/The
Image Bank of Canada:* Photo
No. 2.
Ed Long: p. 13, p. 69, p. 131;
Photo No. 7.
NFB Photothèque: p. 21; Photos
No. 8, 9, 15, 18, 20, 21, 28, 29,
30, 36, 38.
Environment Canada: Photos
No. 14, 24, 25, 27, 41.

CONTENTS

The CBC television program *Arctic Oil* was produced as part of the series *The Nature of Things* by the following people:
Producer and Director: *James Murray*; Writer and Narrator: *John Livingston*; Photography: *Rudolf Kovanic*; Editor: *Michael Bennett*; Sound Recording: *John Crawford, Gerry King*; Sound Editor: *Clark Hill*; Lighting: *Archie Kay*; Assistant Cameraman: *Neville Ottey*; Sound Mix: *Joe Grimaldi*; Music Consultant: *Patrick Russell*; Film Quality Control: *Jim Y.C. Lo*; Cameraman Kurdistan Film: *Fred Macdonald*; Film Research: *Unita Williams*; Production Assistants: *Kay Nagao, Lars Isaksson*; Research: *Juliet Mannock*; Unit Manager: *George McAfee*; Scientific Advisor: *David Nettleship*, Canadian Wildlife Service.
Acknowledgements: Canadian Wildlife Service; Bathurst Island Station, The National Museum of Natural Sciences; Polar Continental Shelf Project; Environmental Emergency Branch, Environment Canada; Petro Canada; The Hamlet Council, Pond Inlet, NWT.

ACKNOWLEDGEMENTS

This book was commissioned by the Canadian Broadcasting Corporation, to expand upon its television program "Arctic Oil", first presented by *The Nature of Things* in December 1979. Acknowledgement is expressed to the following, some of whose remarks in interviews have been quoted here: Titus Allooloo, Max Dunbar, Jake Epp, Donald Gamble, Gerry Glazier, William Henry, David Nettleship, Carson Templeton, and Andrew Thompson. Two others who appeared but are not quoted here from the program are Thomas Berger and John Fraser. David Suzuki conducted the interviews and Juliet Mannock provided the research for the program, which was conceived and executed by James Murray, Executive Producer. For those who would like to follow it up, material from my script for that telecast is to be found chiefly in Part Two.

I am grateful to both Carolyn Dodds and Ingrid Cook for their supportive editorial assistance, and to the Canadian Arctic Resources Committee for pertinent background material on the issues. Also, several of my graduate students and colleagues—especially Neil Evernden and William Leiss—have made both conscious and unwitting contributions to my comprehension of the problems through their work in the Faculty of Environmental Studies, York University. Apart from those attributed to individual sources, however, the views expressed here are mine, and certainly not necessarily those of the Canadian Broadcasting Corporation or any members of its staff.

Although no blame for my opinions, for my interpretation of the evidence, or for any of the other shortcomings of this book may be attached to their names, the inspiration provided in other days, in other places, by Tom Barry, Doug Pimlott and Ray Schweinsburg was always close at hand in the writing.

John A. Livingston
Sunderland, Ontario
1980

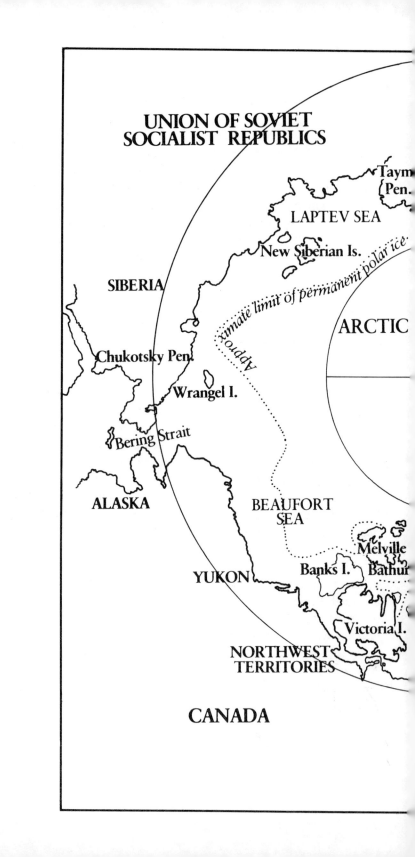

THE ARCTIC

KARA SEA

Novaya Zemlya

BARENTS SEA

FINLAND

SWEDEN

FRANZ
JOSEF
LAND

NORWAY

OCEAN

Spitzbergen

NORTH SEA

North
Pole

Arctic
Circle

GREENLAND

ICELAND

Ellesmere
I.

Devon I.

Lancaster Sound

Baffin Bay

ATLANTIC
OCEAN

Baffin I.

Davis
Strait

Foxe
Basin

LABRADOR SEA

Hudson
Bay

Ungava Pen.

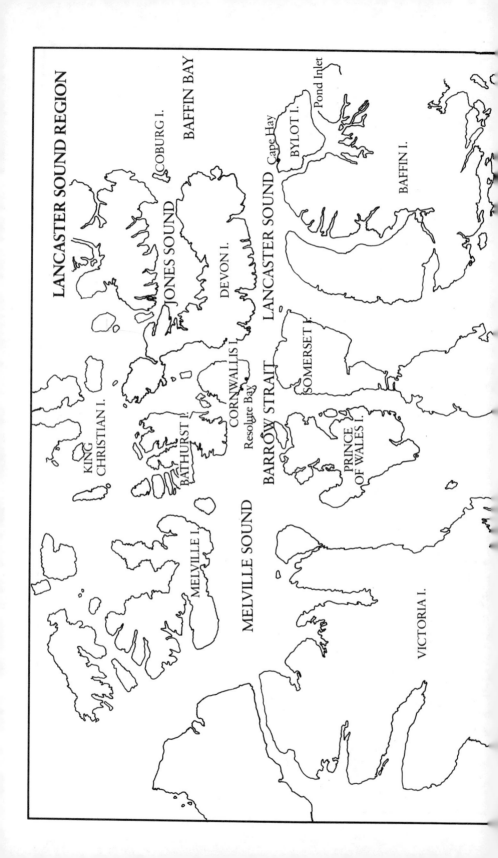

UKON NORTH SLOPE

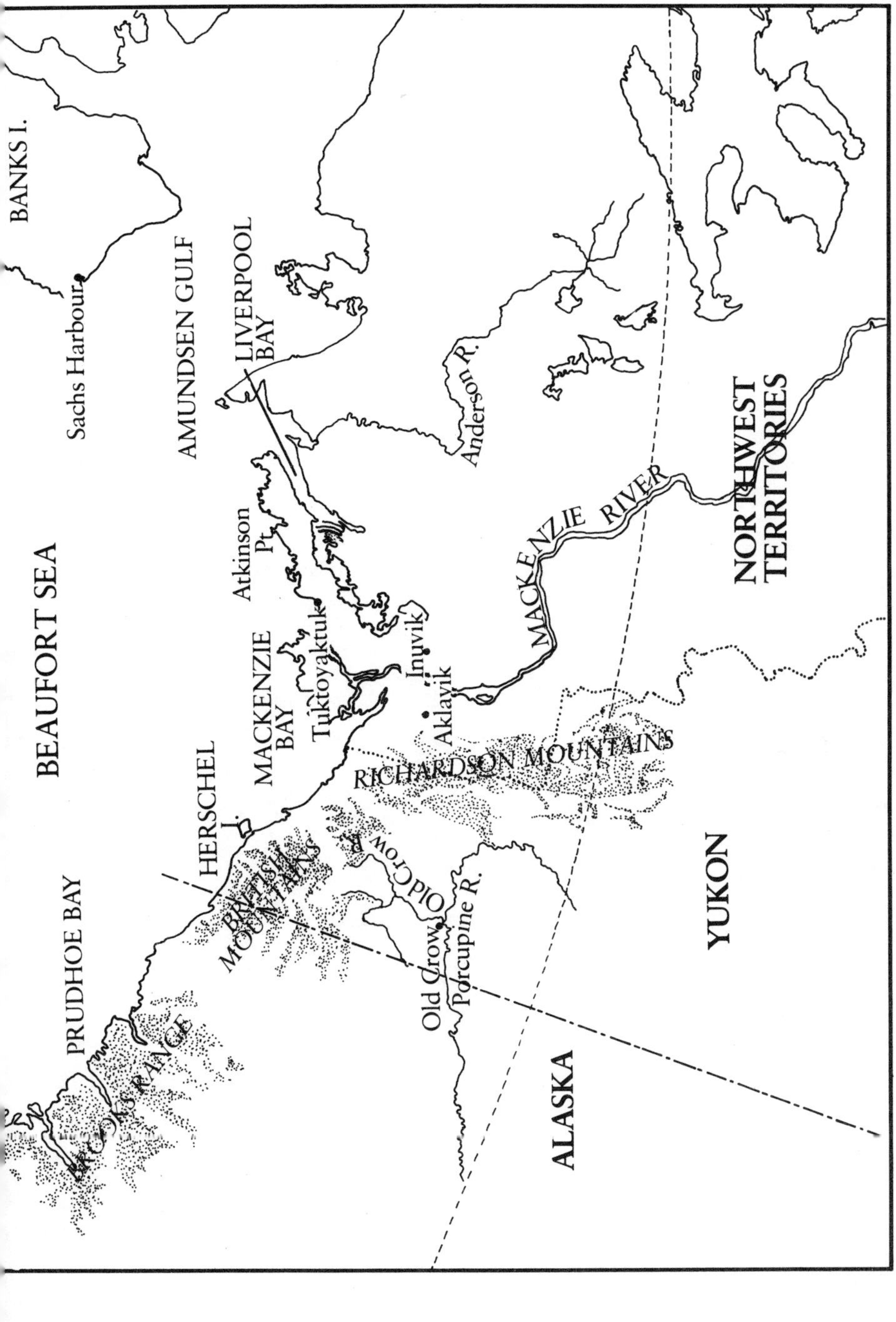
BANKS I.
Sachs Harbour
AMUNDSEN GULF
LIVERPOOL BAY
Anderson R.
BEAUFORT SEA
Atkinson Pt.
MACKENZIE RIVER
NORTHWEST TERRITORIES
MACKENZIE BAY
Tuktoyaktuk
Inuvik
Aklavik
RICHARDSON MOUNTAINS
HERSCHEL I.
BRITISH MOUNTAINS
Old Crow R.
Porcupine R.
Old Crow
YUKON
PRUDHOE BAY
BROOKS RANGE
ALASKA

INTRODUCTION TO THE ARCTIC "DEBATE"

Arctic settlement, Frobisher Bay.

The fate of the Canadian arctic has for a good many years, is now, and for the foreseeable future will be, in dispute. The future of the last large "frontier" region of Canada has become of pressing importance not only to those who have always lived there but also to southern bureaucrats, politicians, "developers", engineers, jurists, scientists, polltakers, technicians, managers, academics, planners, resource conservationists, environmentalists, nature protectionists, analysts, recreationists, entrepreneurs, constitutionalists, consultants, writers of letters to the editor, other writers, pundits, film-makers, ideologues, investors, journalists, commentators, bleeding hearts, troublemakers, naturalists, opportunists, and various others. There is even a goodly number of those who are furious and resentful about all the fuss. Each of the "constituents"—if only because of (even unilateral) individual involvement—has a very real vested interest in what is going to happen in the north. But they all have one thing in common. They consist of both northerners and southerners, but they are all people.

There are still others, however, with legitimate vested interests who neither voice their views nor hear ours. On the admittedly preposterous assumption that a walrus were to achieve standing before some tribunal, no doubt he would have many things to say on behalf of the inarticulate classless clams and mussels of the ocean bed. A shrimplike crustacean called an amphipod might rise on the part of the minute squidlike copepods he depends upon, and the seabirds of Lancaster Sound would have a compelling case for the helpless amphipods. Ivory gulls would plead for their benefactors the polar bears, bears for seals, seals for arctic cod. Foxes would argue for lemmings, and lemmings for grasses. Caribou would be represented by wolves, and wolves by ravens, eagles, gulls and jaegers. Grizzlies would fight for ground squirrels; snow geese would speak for sedge meadows, which would in turn explicate their vested interest in the sun and the rivers and the permafrost itself.

Arctic beings and processes are not of course voiceless; we simply choose not to hear them. We elect not to recognize them. Complicated as the public discussion over northern policy has become, it has not as yet entertained the additional complication that would arise were it to become anything more than a unidimensional proceeding, with any more than one interest represented. There is after all only one protagonist, and he is talking to himself. The arctic "debate" is not a debate at all; it is a monologue. The single participant is ourselves. Were the implications and the possible outcome not so tragic, the whole charade could be dismissed as mere absurdity. But of course absurdity is never "mere"; it can be dangerous.

It is the absurdity of the arctic "debate", and the frightening absence in decision-making councils of any dimension or factor other than the urban-industrial self-interest that impels some people to make films and write books about it. What the northern discussion requires is more, not less, public attention. No possible approach to building and maintaining that attention should be ignored. There are as many ways of doing that as there are participants, but as in most public discussions, positions on arctic affairs tend to cluster around a few points of view. For example, many people are presently working in the field of energy policy; many others are interested chiefly in the refinement of technology toward a safer and more environmentally appropriate approach to resource extraction; still others are primarily concerned with the economic and social implications of northern ventures.

That these are important aspects of the northern question there can be no doubt. However, just as all of the constituents in the arctic discussion are people, so too the major or most common approaches to the present dilemma all are ways and means of adjusting or fine-tuning the process of industrialization that is the overwhelming "thrust" of our society. Most criticism of arctic issues tends to take as given the direction of contemporary human affairs, and concentrates on questions of how to conduct ourselves with greater sensitivity, awareness, and appropriateness. Such criticism tends to be packaged in the short- and long-term interest of the society we live in. Thus the questioning process is severely limited; where solutions are proposed, those solutions tend to be dependent on asking old questions in new ways. For example, we ask how to carry on arctic exploration in safety and in environmentally appropriate fashion. We rarely ask how *not* to carry on arctic exploration, because that is a *new* question and we are uncomfortable with it. It does not fit with our basic assumptions about our society, our culture, and our civilization.

Like it or not, however, serious consideration of all of the implications of northern affairs must lead us to such questions. When we get around to acknowledging this quite simple fact to ourselves, we begin to appreciate the enormity of the stakes that are involved. If we really intend—whatever our reasons—to try to actually *save* the arctic, then the task turns out to be much greater than the mere shaking-up of government, bureaucracy and industry. It means an internal shaking-up of our most fundamental beliefs about nature and about ourselves. Of course there is little we can do about those beliefs until we are able to understand them for what they are. The best way of going about that is by looking at the way in which we see things, because our perceptions are, after all, the products of our beliefs. This book is meant to help the reader approach these tasks.

Although many of us tend to perceive a stereotyped arctic—one unbroken and homogeneous vast expanse of rock and ice, the fact is that there is no single "arctic" about which anything of much value can be said, beyond the most sweeping generalizations. At the simplest level there are two arctics: the marine arctic and the land arctic. But even this distinction is untenable when we take note of the polar bears and seals and seabirds and of the marine invertebrates and fishes that support them.

It turns out, really, that everybody has a different set of definitions according to his specialty. A marine biologist sees the arctic differently than a botanist does, for example. For purposes here, however, let us assume that there are three gross categories or zones, and that they include both land and water. The "high arctic" is the northern archipelago and its surrounding seas. The "low arctic" is the mainland tundra, including the coastal ocean, south to the limit of trees. The "subarctic" is the zone of tension between the tundra and the spruce forest—the land of "little sticks" and muskeg as epitomized by the scraggly spruces around Inuvik or Churchill, and across the great continental barrens.

Also, as we consider the natural history of the north, we must try to erase the map of Canada (especially a Mercator projection) from our minds. To properly "see" the arctic, we must look down on a globe from above the North Pole. A bull's eye of concentric circles emerges: the roughly circular crown of perpetual polar ice, surrounded by a wide necklace of land and water that becomes free of snow and ice in summer, and finally the great circumpolar robe of dark green spruce forest into which the tundra eventually disappears. Keep in mind that *the arctic is a northern phenomenon, not an American or Asian or European one.* This has recently been illustrated by the Inuit Circumpolar Conference, which from an environmental point of view makes a great deal more sense than do the sundry names of the assorted nation states which penetrate the north.

Special attention will be given to two selected critical areas in the Canadian north. Both are biological "oases", fertile areas in the arctic desert, and both are threatened by industrial invasion. One, the low arctic Yukon North Slope, was recommended in the report of the Mackenzie Valley Pipeline Inquiry in 1977 as a National Wilderness Park. To no one's genuine surprise, that recommendation has not yet been acted upon by government. Indeed the entire Yukon and Northwest Territories coast from the Alaska border to the Amundsen Gulf, and including the delta of the Mackenzie, is grievously endangered today because of offshore drilling in the Beaufort Sea and its attendant pressures—such as the seaport and tundra pipeline currently proposed by Dome Petroleum. No environmental hearings have ever been required for Beaufort exploration. In fact, although the approval was given to Dome during the Berger Inquiry in 1974, drilling had in fact been sanctioned by the Cabinet the year before, at a time when Mr. Justice Thomas R. Berger had not even begun his inquiry. Thus, the north coast, particularly that of the Yukon, is one area of special interest.

The other, the region of Lancaster Sound in the high arctic, was the geographic focus of the film "Arctic Oil" as presented by *The Nature of Things*. This is another outstanding oasis in the biological sense, and it is also of consuming importance to government and industry for offshore drilling and other exploration, as well as being part of a proposed route for year-round tanker traffic. The most advanced proposal has tankers carrying liquefied natural gas to eastern Canada from Melville Island on the northwest rim of the high arctic archipelago, by way of Lancaster Sound.

17

Part One of this book deals broadly with the natural history of these two "target areas" in particular, by way of a few selected examples of how plants and animals go about their daily and seasonal affairs. Although conditions are never precisely the same in any two places in the arctic, we may in this way gain a general feeling for the nature of life process in this extraordinary land. The next two parts of this book, again by way of example rather than by any attempted comprehensiveness, but wholly in the contemporary spirit that nothing is impossible in the north, will attempt to "section" an iceberg.

The visible, above-surface part of the iceberg is used to depict the "issues"; the invisible, sub-surface part, the "problems". This distinction is important. The issues are clearly to be seen and read about. They are illustrated in all dimensions every day, complete with antagonists, controversy, debate. The problems, on the other hand, are rarely perceived and even more rarely addressed. This is not to belittle the issues, of course, because the future integrity of the arctic depends on their resolution. But that resolution is quite probably dependent ultimately on our willingness to acknowledge and act upon the problems.

For example, pesticide pollution is a public issue, and has been a hot one since well before *Silent Spring* was published in 1962. The (relatively obvious) underlying problem, as distinct from the issue, is a society that can condone the use of poisons to advance its aims. Buried still deeper are the cultural roots of the society itself, which not only make possible the condonation of toxic warfare on our living habitat, but also provide elaborate rationalizations for our unique behaviour. One such root is the generally unchallenged cultural assumption that human (especially Western industrial) well-being comes first, and that if other sensate organisms must be sacrificed in order to maintain and enhance that well-being, then that is in the natural order of things. It is *necessary*.

Another example of "sacrifice" to humankind is the unspeakable tortures to which non-human beings are subjected in the course of the advancement of medical technology. The "issue" is the familiar vivisection discussion. The *problem* is the set of cultural assumptions that sustains the sanctity and blessedness of medical technology—at *any* cost. The agony of individual sentient beings,

18

deliberately inflicted and often hideously prolonged, is seen as a necessary cost, an "externality". The human purpose is the cosmic purpose.

We shall look at the arctic "debate" by way of both issues and problems. The issues in Part Two are familiar: the much discussed industry-government "conspiracy", the bungling of the bureaucracy, the scandalous subsidy of the multinational corporations by the Canadian taxpayer, the hazards of oil spills in arctic ice, ecological disruptions and possible collapses, native claims, protected zones, economic and social impacts, the environmental assessment review process, and so on. All of these are issues involving combatants; many of them have become "high profile". The tip of the iceberg is brightly lit and in high relief, for anyone to see who cares to look at it.

The rest of the iceberg, however, the greater part that slopes down and away, beyond the murky limit of visibility, is infinitely more subtle. We tend to not even acknowledge its presence, much less go for the scuba gear to see what actually may be down there. It might be terrifying, for here we encounter the strange and often unthinkable buttresses and bulwarks that give rise to and support the issues. The iceberg is not floating free, as it turns out. We find it grounded fast to a bedrock of pervasive social beliefs and assumptions. At last we face the problems.

Much of the energy invested in chasing issues could well be applied to problems. Part Three will address, among others, such problems as the morbid fear of alternative perceptions that is part of our society and culture. That fear has both historical and biological roots. There are historical and cultural reasons for the peculiar way that southern "civilized" folk see the north and its inhabitants, including its original human inhabitants. There are parallel reasons for our concepts of "resources", and of "our arctic heritage".

Perceptions govern almost all of our actions. If we apprehend things in a certain way (which may turn out to be quite arbitrary), our behaviour follows. Various "constituencies" see arctic issues (and occasionally problems) in various ways, but those ways have many common denominators. Shared perceptions are the base of the iceberg, solid frozen to the bedrock.

But it is, after all, only frozen. Times change, and thaws do

19

happen. The history of this planet is a chronology of changes, including freezes and thaws. Some have been dramatic, most less so. Part Four will attempt to present some optional changes. For those who wish to perceive them, they are sometimes called "choices".

A BRIEF HISTORY OF OIL, GAS & ICE

Inuit seal-hunters.

The "northern vision" of Diefenbaker days was well established and government and industry had been gearing up
for a decade, with exploration well advanced, when word
exploded in the business press of the discovery of oil and gas in large
quantity at Prudhoe Bay on the Alaskan coast in 1968. At the time,
the implications of the discovery may not have been clear to most of
us, but not for very long. There were subsequent discoveries in both
the Yukon and Northwest Territories—at Atkinson Point on the
Tuktoyaktuk Peninsula, and in the Mackenzie Delta, to be followed
by additional finds in the high arctic islands and, most recently,
offshore in the arm of the Arctic Ocean known as the Beaufort Sea.

What was at the turn of that decade little more than old-
fashioned dog-eat-dog commercial competitiveness was transformed gradually by political impetus from Ottawa, and then
dramatically by the Organization of Petroleum Exporting Countries (OPEC) and the "energy crisis" into a flailing, jostling, no-holds-
barred stampede. Fuel was added to the rocketing acceleration of
the "development" imperative by growing unease about global oil
supplies. The geopolitical circumstances surrounding fossil fuels in
the middle and late seventies allowed the arctic "development"
imperative, in the perceptions of the crusaders in the multinational
oil companies and a succession of federal governments, to become a
mission as sacred and unchallengeable, as "fundamentalist" in its
self-righteousness, as that of any Ayatollah. The arctic "development ethic" had of course been in place in Ottawa since the 1950s.

As early as 1968 the government of Canada had struck a "task
force" on northern development, and the National Energy Board
had opened discussions with private corporations, whose unquestioningly acknowledged expertise in fossil fuel extraction and
marketing equipped them to prepare national goals and strategies,
including policies, better than any independent group the government could muster. In 1970 a set of federal "guidelines" for the
building of northern pipelines was issued—in the total absence of
public consultation in any part of the country. A pipeline or
pipelines was taken as given. The Inuit were entirely ignored.
Alarmed, that year the Inuit in the Delta Region formed the
Committee for Original Peoples' Entitlement (COPE), and in 1971
they joined together with the Inuit from across the north to form the

Inuit Tapirisat of Canada. The following year, gas having been located in the Mackenzie Delta, the federal government issued ludicrously elastic and incredibly imprecise "expanded" guidelines for pipeline building. Desperate to head off the shipment of Prudhoe gas to the lower United States by way of Alaska, the Canadian government and various (forever shuffling and reshuffling) consortia of oil and engineering firms rushed hell-for-leather to prepare a Mackenzie Valley alternative. A pipeline construction proposal emerged. Much as the 1970 Arctic Waters Pollution Prevention Act had been designed not to inhibit industrial activity by way of environmental controls but rather to assert Canadian sovereignty in the arctic seas at a delicate time, the Mackenzie pipeline push had little if anything to do with Canadian energy needs and nothing whatever to do with the desires or wishes of the Canadian public, including the Inuit.

As early as 1969, numerous individuals, a few conservation organizations and elements of the press had voiced sporadic and totally ineffective protests (having been among them, one can say that). (Conservationist James Woodford later reported that "by the end of April 1970, the federal government had issued about 9,450 permits and 445 leases on 456.7 million acres in the Northwest Territories and the Yukon." Without public consultation.) In the winter of 1970–71 there began to be apparent a pressing need for public information and participation in a process of policy- and decision-making that was going forward completely independently of public opinion. The government-corporate juggernaut had achieved its own autonomous momentum. One individual decided to do something about it. The tireless designer and prime mover of an instrument to meet the obvious need was the late zoologist Douglas Pimlott, who with Richard Passmore and Kitson Vincent founded an *ad hoc* citizens' information group they christened "Canadian Arctic Resources Committee" (CARC). A larger number of concerned and distinguished Canadians soon coalesced around that original nucleus. Never had a public watchdog been needed so urgently.

It was also in 1971, by the way, that *The Nature of Things* presented its first film "special" on the far north, "The Living Arctic". This award-winning production was shown at CARC's first public conference, the (having in mind the temper of the time) extraordinarily successful "National Workshop on People, Resources and the Environment North of 60°", held in Ottawa in the spring of 1972. The proceedings of that first open, uninhibited, public assessment of northern affairs were published under the title *Arctic Alternatives* the following year. Although there have been many crucial revelations since that time, the book remains a bench mark for social, political and biological perspectives on the Canadian arctic.

CARC enjoyed instant recognition in terms of national interest and (at least moral if not material) support. At last the public attention began to be directed in a systematic and significant way to what was actually happening in the north—or, more correctly, to what was actually happening in Ottawa and how it was affecting the north. 1972 would witness both the growing acceptance of CARC and a federal election. The result of that election was a Liberal minority government which, as such governments are wont to be, was soon at pains to reveal a highly developed social conscience. One manifestation of this was the announcement at the beginning of 1973 that there would be public hearings on the Mackenzie pipeline proposals. As political scientist Edgar Dosman describes it, "hearings, however unpalatable, were essential to restore confidence, to show that Ottawa was genuinely concerned about native people." And of course about questions of the biophysical environment.

Thomas Berger, the man picked to conduct the Mackenzie Valley Pipeline Inquiry, understood the mandate of his commission was ". . . to consider the social, environmental and economic impact of a gas pipeline and an energy corridor across our northern territories. . . " Two formerly competing consortia had by this time merged into one proponent, Canadian Arctic Gas Limited. Both had been doing impact studies in anticipation of that requirement.

In spite however of the Berger hearings, then in progress, and in spite of the perhaps wishful public conclusion that at last northern happenings were going to be held up to open scrutiny, in 1974 the federal government granted Dome Petroleum approval to do exploratory drilling offshore in the Beaufort Sea. Drilling began in 1976 and continued through 1980 without any assessment of the environmental hazards ever having been required.

24

There are many horror stories—more indeed than the ordinary
mortal might think possible. They will not be recounted here;
thanks chiefly to CARC and to its friends and allies there is ample
documentation. There is no longer any excuse for any of us who feel
we have a stake, or want to participate, not to be up to date. The
watchdog is watching—constantly.

Issues of the day do not spring forth full-blown out of nowhere.
All have a genesis and an evolution. All develop out of a "mix" of
political, economic, social and cultural influences. Events have both
causes and results, and that is why we read history. The causes are
complex and the results are not predictable. The history of the
present situation in the arctic is essential not only to our understanding of how we managed to get ourselves where we are, but also to
whatever deliberations we may undertake on what if anything to do
now. This book is filled with the names of those who have
attempted to deal with some of the questions relevant to our
discussion. The reader may wish to refer to the actual publications,
which are given in the Selected References.

As an antidote to the depressing and downright frightening
history of the development of Canadian public policy for the north
(described by Pimlott, Brown and Sam as "a story of the ascendancy
of bureaucracy over democratic process and of secrecy over public
participation and involvement"), we have the refreshing forthrightness, clean reflection and uncluttered vision of Thomas Berger. His
report on the Mackenzie Valley Pipeline Inquiry, published in 1977
under the title *Northern Frontier, Northern Homeland*, provides
the background, evidence and reasoning behind his decision to
recommend that no pipeline be built across the northern Yukon and
that a Mackenzie Valley line be postponed for ten years. (You
always seem to remember where you were when "big news" hit.
After having been involved in the Mackenzie Valley proceedings in
various ways, I learned of the judge's decision from a battered and
filthy copy of *Time* magazine in a somewhat disreputable shop in
Kuala Lumpur.)

Berger was especially concerned that the settlement of native
claims should precede "development", but he also demonstrated
rare sensitivity to the unique environmental conditions of the north
and the delicacy of its natural communities. His report is a solid

basic introduction to the total environments of the western mainland low arctic and subarctic. It is also very pleasant reading.
Considering its deep understanding of the relationships between
human beings and their natural environments, I was both com-
forted and encouraged to find Berger's report shelved under "Humanities" in the York University bookstore. Humanities don't
usually acknowledge our biology.

Many people took heart from the Mackenzie Valley Pipeline
Inquiry and its outcome. In *The National Interest,* however,
Dosman brought us abruptly back to earth by describing the
Commission as "an accident, an anomaly, which no future Liberal
Government will entertain again. Born of a minority Government,
with Prime Minister Trudeau on the defensive, it was the product of
unique circumstances." No doubt it was. One silver lining is that
Berger shook the government-bureaucracy-corporate establishment
very severely indeed. More important, he showed us that there are
always optional ways of looking at things. The conventional
"development" wisdom is not necessarily *the* wisdom, Berger is
telling us. For example, the essence of the environmental assessment
process, from the point of view of a proponent, is fragmentation.
Canadian Arctic Gas, the proponent of the Mackenzie Valley gas
pipeline, actually attempted to have the potential impact of the
pipeline reviewed in isolation from the impact of the processing
plants that would deliver the gas to the pipe. The review was to be of
a pipeline magically set into place without having any connection or
relationship with anything else. A proponent will always present the
pieces, not the whole.

Berger was not having that. He recognized that the north is a
continuing process of dynamic relationships, not an aggregate of
static parts. In the course of his investigation he took the long view,
not the short; the regional view, not the local; the individual view,
not the institutional; the human view, not the organizational. In so
doing, whether or not such an inquiry is ever repeated in quite the
same way, Berger broke a brave wide trail that many in industry and
government would happily forget. Berger must be remembered and
emulated, and the bureaucratic-corporate "developers" must be
reminded of him through the eighties and beyond.

Since the final report of the Mackenzie Valley Inquiry came

26

down in 1977, the activity of government and industry has at least temporarily shifted away from the Yukon mainland onto the continental shelf of the adjacent Beaufort Sea. Also, the longtime enthusiasm for the riches of the high arctic islands has intensified. Pressures on both offshore and land areas are building daily, everywhere in the north.

Will living beings—plants and animals—of high latitudes be able to live with the industrial search for arctic oil and gas? Will they be able to so much as survive it? As you will see in Part One, what little understanding we do have of the nature of life in the north gives us ample cause for misgiving.

PART ONE
ABOUT COPING

Snow goose, Devon Island.

Everywhere on this planet, existence for any living being is a matter of coping. Life for any individual plant or animal is a series of unpredictable events surrounding the satisfaction of primal needs—food and drink, growth and reproduction. There are as many ways of meeting those basic requirements as there are species of sentient beings—literally millions. It is a fundamental assumption of biology that no two species, even though they may be closely related and may live in essentially the same kind of environment, make their living in *exactly* the same way. Every one copes slightly differently. This is what makes work for ecologists.

World environments vary in the opportunities they offer. A tropical rainforest, for example, contains so many "micro" environments, from the leaf litter of the floor through rotting logs and shrub layers to the topmost leafy canopy, that there are "niches" (ways of coping) for a great many species of plants and animals. Even deserts have a surprisingly wide assortment of places for those who are adaptable enough to take advantage of them. In this respect arctic regions are the opposite of the tropics: there are fewer niches, as evidenced by the relatively fewer kinds of plants and animals able to deal with a smaller variety of opportunities.

But there *are* ways of coping in the arctic; simply witness the fascinating assembly of those who do. If they couldn't take it, they wouldn't be there. But they are there, and from their continuing survival in what might seem to us to be such a forbidding land, we can properly infer that in the far north there are many and some quite interesting ways of dealing with the changing circumstances of daily and seasonal living.

The limitations to life in the arctic are severe, as everyone knows. In general, except for a few brief weeks in summer, it is simply too cold for most living beings. With local exceptions, there is very little rain (or snow); indeed much of the high arctic qualifies in this sense at least as a proper desert. Even where there is fresh water lying about, the air is so cool that there is very little evaporation. In spite of a glorious but fleeting period of perpetual daylight, the total amount of solar energy available is very low. Some places are almost constantly overcast, even in summer. Also, as we have seen, there are fewer plant and animal species, and thus fewer biological communities, than in most other parts of the world. The arctic, as the ecologists would put it, lacks "diversity".

Such general conditions are sufficiently sombre for a start, but when you scratch the surface a little deeper, you begin to appreciate the fuller enormity of arctic bleakness. There is little soil in the far north. It tends to occur in pockets, and even there it is usually thin and very poorly drained. The growing season is so short, and the soil base so limited, that plants grow (in the sense of spreading) slowly, which also means that they *evolve* slowly. Change in the evolutionary sense, which is adaptation to changing environments, is dependent on rapid turnover of many generations. Vegetative life is too laboured in the north for the process of adaptation to proceed as rapidly as it does elsewhere.

All of these factors add up to the fairly obvious and quite common conclusion that the arctic has an unusually low "carrying capacity" for all forms of life. It simply does not have enough energy or nutrients in its base to sustain very many different forms of life, or significant populations of those forms that are there. In a very general sense, on the land at least, this seems to be true. Based on the depressing litany of constraints to biological process, one might quite reasonably conclude that the arctic is poor and "primitive", that it is so limited in so many ways, so lacking in the kind of elasticity that more abundant environments enjoy, that its built-in resistance to sudden change of any kind must be much lower than that of "richer" areas of the world. Perhaps the arctic is indeed brittle, and its life structure could snap without warning because of some unexpected pressure. All of this may well be so—but the reasons may have nothing to do with the arctic's low "diversity" and "carrying capacity".

You see, when we look at the marvelous and beautiful collection of beings, from saxifrages to snow geese, from poppies to polar bears, from corals to caribou, that *do* cope (and most successfully) in the north, then perhaps conditions may not be quite so inhospitable as we have been led to believe. When we think too of the Inuit who over unknown thousands of years developed a culture so sensitive and decent, of such wisdom and ingenuity, in such apparently forbidding circumstances, it leads us to wonder. The

31

arctic has in fact produced things of a beauty—indeed grandeur—
unsurpassed anywhere on Earth. Could such a crucible have "low
carrying capacity"?

Let us look at a few examples of how some life forms in the
north manage to handle themselves in their "hostile" surroundings,
how the arctic seasonally burgeons and explodes with hot-pumping
life in all of its "austerity". We cannot begin merely at the grass
roots, for that would be to overlook the most fundamental factor
underlying all arctic terrestrial life, the permafrost.

THE ARCTIC PERMAFROST

Permafrost is ground that is permanently frozen, year in and year
out. It is not necessarily frost; it may be sand, or gravel, or mud, or
solid rock, so long as it stays below 0°C perennially. Sometimes it is
pure ice. There is a great deal of permafrost in the world—about 20
per cent of the earth's surface is underlain by it—and in some places
it is hundreds of meters deep. But it isn't static. It forms—and
grows—in places where more heat is lost to the atmosphere each
year than is gained; in other words, what the economists would call
a heat "deficit". This applies to most of the terrestrial arctic, except
under large lakes, rivers and bays, where the water has an insulating
effect, acting as a thermal cushion. Because it is a living and growing
thing, it is no more permanent than it is frost, but for any of our
purposes it might just as well be. Significant change in permafrost
distribution is measured in geophysical time, not our time.

There are however two sorts of permafrost. One, the deep
layer, is as described above. At the surface there is a second zone that
the warmth of the sun manages to reach each summer, and it is this
"active" layer—annually thawing and freezing—that sustains all life
on the arctic land. Were it not to thaw each year, no plants could

---

1 *Polar bear near Devon Island, Lancaster Sound.*

2

2 *Snowy owl populations depend on populations of lemmings.*

3 *Arctic poppies thriving around muskox skeleton, Bathurst Island.*

4 *Peary's caribou on the summer tundra.*

3

4

5 *Icebergs off Coburg Island,
Lancaster Sound.*

6, 7 *Lichens grow wherever there is
the slightest opportunity—on the
warm rock surfaces, in the shelter of
a decomposing log.*

7

8

9

8 *Once the Inuit hunted the walrus armed only with harpoons made of wood, bone and ivory.*

9 *A walrus may be over 900 kg and its tough hide will dull a knife.*

10 *Huge numbers of murres colonize the cliffs of Coburg Island.*

11 *Male willow ptarmigan in spring change of plumage, Yukon North Slope.*

grow, and nothing further could happen in a biological sense. This annual thaw may only penetrate for a very few centimetres, but where it does it is sufficient for plant roots to do what they must do during the precious eight weeks or so of growing season.

In permafrost we find the secret to the peculiarities of northern soil drainage. Melting snow, and such rain as falls, tends to move over the land in sheets or wide streams, because it cannot percolate downward. But the land surface is uneven, and some parts of the arctic consist of thousands of large and small lakes, tarns, pools and puddles—fresh water with nowhere to go but to collect in depressions until the autumn freeze-up. (Since it is the permafrost that keeps the water at the surface, where plants can get at it, we see that permafrost is not all bad; we have in the relationship between plants, soil and permafrost a kind of symbiotic paradox.) For all the same physical reasons, much of the tundra is covered by soggy meadows. It is this vast store of standing fresh water that offers such abundant opportunities for animal life—mosquitos, for example, once the temperature of the air reaches about 12°C.

THE PLANTS

Under the constant spring and summer sun, permafrost would thaw much more deeply than it does were it not for an insulating cover of vegetation. For reasons we have seen above, we might see this as a kind of "enlightened self-interest" on the part of the plants. Where vegetation is removed or disturbed (sometimes by natural mudslides, but most often by human activity) the resultant penetration of warmth, and thus of thaw, more deeply than usual into what was the integrity of the permafrost can cause serious damage. Quickly there appears a mud puddle that extends itself down into the permafrost. Unless the vegetation is quickly replaced (arctic plants spread slowly), the open sore may widen, expand and deepen, year by year. This is the effect we see after wheeled or tracked vehicles have moved over the tundra, breaking the insulating surface. The tracks become furrows, often with gushing erosive streams, which cannot heal until vegetation can recover them or until some theoretical balance of heat and cold is achieved. Such scars are today everywhere apparent in the wake of geological crews and other exploratory activity. Much of the western arctic and subarctic looks as though someone had gone at the Mona Lisa with straight-edge and felt pen.

Where the vegetation is intact, it proceeds in its own deliberate way. The thawed active layer, though perhaps very shallow, delivers sufficient moisture and nutrients for the year's life processes to go forward. Arctic soils, what there are of them, are usually low in nourishment, especially nitrogen, mostly because bacteria do not fare well in these short seasons of low temperature. There is an immediate joyful response by the plants, however, in places where some animal has come to final rest, and whose remains slowly break down to rejoin and thus enrich the soil. A cluster of gay and brilliant flowers will appear where some muskox, caribou or other carcass has eluded wolves and ravens—a cameo work of art in illustration of the ineffable wholeness of life process. There is nothing quite so exquisite—anywhere. Arctic fox dens, which are so often finely— almost outrageously, sometimes—festooned in multicoloured nose-gays are also remarkable sights. There is so little soil in the north that is genuinely diggable that a good den site may be used by fox generations for scores, perhaps even hundreds, of years. What minute scraps and shards of debris may escape sharp eyes and noses make superb compost, over time.

The rigidity of the growth regime has important evolutionary implications for arctic plants, many of which have become dwarfed versions of perennial flowers, trees and shrubs we know well in the south. Few plants enjoy constant wind; it dries, shreds and batters them. So tundra plants lie low, hugging the surface of the ground, thereby also benefiting from the shallow "heat bounce" at that level. An often and uncritically repeated definition of the arctic hinges on the absence of trees, which simply isn't accurate. Much of the tundra is covered with trees—willows, for example—but instead of boasting the great sturdy trunks and cascading limbs of their southern relatives, arctic willows grow flatiy along the ground, pressed against it vinelike, as though clinging with desperation for whatever miserly mercies the land may afford. But that is appearance only. The "trees" do very well indeed, and live long. A willow with a bole only an inch or so across may be two or three hundred years old, perhaps more.

Where there is shelter from the wind, shrubs can venture somewhat farther from the ground, and such is the case in favourable valleys where more recognizable birches and willows may thrive in profusion, several feet tall. In these sheltered "sanctuaries" the leaf litter can begin to build up a kind of basic topsoil. The plants can gain a stronger and more nutritive roothold. Such places provide cover, food and nesting sites for small tundra songbirds such as redpolls. In the Yukon in such situations there are even yellow wagtails from Eurasia.

But there are other, more important evolutionary side effects than mere form and structure. The shortness of the growing season in the far north, together with the vagaries of the weather, mean that in any given year there just may not be time for "normal" plant processes to take place. Flowering plants, like any other beings, depend for evolutionary change on sexual reproduction—the mixing of the genetic material of two individuals which results in offspring slightly different from either of the parents. When one of these slightly different new individuals has something about it that allows it to prosper slightly better in a given (perhaps changing) situation than its antecedents did, live longer, have more offspring, and so on, we call that change "adaptive". Adaptation is a change that is perpetuated by inheritance and that allows a species to go on "doing its thing", or a slightly different thing, a little more efficiently than before. It goes on all the time, in all living beings. It has to, because environments are always changing. But adaptation depends on the mixing and remixing of genetic material. Sometimes the arctic summer is so short that there is not sufficient time for the plants to reproduce by flowering and seeding, so they may reproduce asexually, by sending out shoots from the central parent plant which will themselves take root and become new plants. This produces the "pincushion" effect so common in arctic flowers.

Of course these "new" plants are merely extensions of the original. In fact they are, for a time at least, literally part of it. No change has taken place; they are exactly the same, genetically, as the original. There can be no adaptation. That is what is meant by the adaptive slowness of arctic plants; as often as not, plants produce carbon copies rather than genetically different offspring. Evolution, at least in this sense, is at a standstill. Were the arctic environment to change more or less quickly, some say, many plants might not be able to keep up with that change.

The same effect can be achieved—somewhat more elegantly than by mere shooting—by a plant pollinating itself, a procedure with which every gardener is familiar. But even here there has still been no genetic exchange. Nothing new has been added, because the same individual plant is doing no more than combining its own male and female cells. The evolutionary or adaptive result is the same.

One consequence of all this is that many arctic plants appear to have remained more or less unchanged for as long as several thousand years. Some authorities see this constancy of form and (apparent) inability to change as a kind of vulnerability: the plants are at a very real dead end, and might be swamped by varieties of southern invaders in the due passage of time over some warming trend.

Perhaps so, but we do have the evidence of the remarkable ability of arctic plants to flourish wherever they are granted even the merest extra crumb of encouragement. The potential seems to be there. Also, what matters in nature is not the next decade, not the next century, not the next epoch, but the *now*. Nothing else. This year, this day, this moment is what counts. That is what coping is all about. In any event, the plants do very well, as we have seen. They are "survivors" of the most admirable kind. Arctic spring sees the instantaneous explosion of birches and willows, glowing lichens, rich grasses, sedges and rushes, and the welter of small flowering herbs—avens, louseworts, cinquefoils, poppies, primroses, butter-cups—the mosses, ferns and horsetails and all the rest the sum of which is a profuse perfection, in the here and now.

Although there are vast arctic areas without apparent plant life of any kind, to offset that seeming sterility there are countless oases large and small. Each pond and stream has its green edge, each frost polygon has its geometric boundary of fresh vegetation, each boulder has some venturesome being, be it only a lichen, in its lee. Each, in its way, is coping; no more can be demanded of them.

THE BIRDS

If the ubiquitous little green man from outer space were one day to descend upon the arctic, wherever he might find himself he would probably first notice a bird. There are many mammals, to be sure, but birds are by far the more visible. The space visitor might be so unlucky as to land in one of the more "godforsaken" stretches of the landscape, but even there he would be bound to see, sooner or later, some slowly cruising gull, jaeger, or raven. Birds of one kind or another are pretty well everywhere. In fact, if you fail to find a gull, jaeger or raven, one of them will almost certainly find *you*. You are obviously an animal, which means that eventually you are going to have to eat something. Or perhaps die. That is all the inducement they need to keep you well in view.

As in any part of the world, birds of the arctic vary a great deal in their density and distribution over the land. Some congregate to nest by the hundreds of thousands; some gather in fall aggregations that are more like clouds than flocks; some take solitary sanctuary on the most remote cliff faces; some are scattered here and there on pebble beaches or on tundra, both wet and dry. Always, there are birds.

Like other animals, and like plants, birds find themselves in greatest profusion in the tropics where, as we have seen, there are more opportunities for varying lifestyles than elsewhere. As you move north or south from equatorial regions, the variety of animals diminishes until you arrive in the highest latitudes, where the number of species is the least.

Of the roughly 520 species of birds in Canada, only about 75 regularly nest north of the limit of trees. Of these, about three-quarters are "holarctic" or "circumpolar"; they live not only in North America but also in arctic Europe and Asia. Their distribution is circular; they surround the crown of ice at the summit of the globe. Predominant among these intercontinental species are the sandpipers and plovers, gulls, jaegers and auks, the loons, some ducks and grebes, and several birds of prey. The balance of Canadian arctic birds tend to be closely related to their opposite species in the Old World. A few are northern-adapted representatives of strictly American families such as blackbirds and warblers.

Considering the pervasiveness of both salt and fresh water in arctic summer, it is no surprise that almost three-quarters of the bird species have a more or less close affinity to water. The largest single group consists of the shorebirds, gulls, terns, auks and their relatives; the next, the swans, geese and ducks. And all four species of loons in the world are here.

Birds survive arctic conditions by drawing on two basic attributes—hot blood and the power of flight. Protected as they are by the unparalleled insulation of their feathers, birds can survive almost any temperature, at either end of the scale, that they can encounter naturally on Earth. They can tolerate northern conditions quite comfortably so long as they have sufficient fuel to feed the internal furnaces that keep them warm. Food in plenty is the key.

It follows that breeding birds tend to cluster in much the same oases as plants do, where there is dependable food for the nesting season. Small finches such as Lapland longspurs feed their young on the extraordinary abundance of small invertebrates—spiders, mites, sucking insects and so on—that bloom among tundra vegetation in summer. Auks gather in incredible masses on vertical sea cliffs over the rich waters of the Arctic Ocean. Geese colonize low vegetated flats where their shoreline food abounds.

The breeding behaviour of the birds is driven by terrible urgency. Everything is timing. The season is so short that the birds have to make use of every available moment in order to nest, lay, incubate, hatch and rear their broods. Here the overwhelming importance of the "now" is manifest. T.W. Barry, of the Canadian Wildlife Service, who knows more about the waterfowl of the western Canadian arctic than all the rest of us put together, has seen female snow geese that have arrived at their nesting grounds in the Anderson River Delta on the Beaufort coast carrying fresh sperm in their oviducts. So great is the urgency that the birds have already mated before even choosing a nesting site. There is no time for leisurely courtship and pairing. Such niceties have been observed much earlier, and the birds have mated while en route northward, on gravel bars and islands in the Mackenzie River. No instant of time is lost. The first rule of coping is to set about the serious business of getting a brood underway without the slightest delay.

The second rule is to get that brood—and themselves—out of there before the onset of cold weather makes food unavailable. There actually is one Canadian bird that has been thought to at least occasionally hibernate (the poor-will of southern British Columbia and Alberta), but no arctic fowl would be advised to try that. So when fuel for their hot blood fails them, the birds have to resort to their wings. And true flight it is—flight from the imminent onset of winter famine.

Some birds drift south only as far as they absolutely have to. Ptarmigan, for example, go only far enough to find a dependable food source. They are followed, usually, by the gyrfalcons, who require a reliable supply of ptarmigan. Snowy and short-eared owls go no farther than they need to in order to keep in touch with the voles and lemmings. In times of severe small mammal shortages they may move well down into the northern United States. Such birds are not migrants in the strict sense; they are opportunists who go where they must go or where they are driven, but they are not "programmed" in the way that regular migrants are. They might best be described as occasional emigrants.

Truly migratory birds always nest in one area and winter in another, year after year. Both areas, though often large, are specific in that they are "traditional", and the birds move between them semi-annually. Bird banding has shown that an individual may return to literally within a few feet of where it was the year before. The birds migrate regularly between these two home bases, regardless of the fact that there may be apparently perfectly suitable nesting or feeding grounds along the way.

The fabled arctic tern is one of these. Although some terns nest south to New England, others spend the summer as far up the globe as 82°N, and arctic terns are known to winter at least as far as 74°S, a distance that is hard to explain on any grounds. Add to this the fact that many of them add a dogleg of thousands of miles by going down the east coast of North America, then across the Atlantic to the west coast of Africa, then south to the seas of Antarctica! It has been suggested that the terns (and some others) may take these seemingly unnecessary voyages because their migratory pattern may have been formed when "Old" and "New" worlds were much closer together geographically than they are today, and that as continental drift widened the Atlantic, they were bound by their allegiance to in some way keep in touch with the ever-receding shores of Africa. Whatever their reasons, the terns do it. Such behaviour would seem to be of a higher order than mere coping. On the other hand, no doubt that is how coping is perceived by an arctic tern.

Birds have different ways of handling the fuel requirements for migration. Some species stop, loiter and feed as they go; others stoke up before they leave and go enormous distances non-stop. Some move over a wide front, drifting southward over the greater part of the continent; some use ancient, well-defined and often quite narrow channels or corridors.

The stress of migration requires a special kind of coping. There are severe physiological demands on a bird's system at this time, and these must be met by fat reserves that the individual builds up prior to or maintains during migration. Some small birds may accumulate fat up to twice their lean body weight just before taking off.

Undoubtedly one of the most spectacular migration phenomena in the north is the fall gathering of the snow geese of the western arctic on the North Slope of the Yukon and Alaska. This is a strip of tundra between the Richardson, British and Brooks series of mountain ranges and the Beaufort Sea, to which we shall return. Birds that have nested or been raised in the great colonies of Banks Island and the Anderson River Delta, and in smaller numbers elsewhere, move westward in autumn by way of the Mackenzie Delta, to spend the last days of the season feasting on nutritious berries and sedges. If they are not disturbed, they will take on sufficient fat there to get them non-stop to their next traditional fueling point in northern Alberta. In some years, however, early blizzards in late September or early October may drive them off the Slope before they are ready for the Alberta leg of their trip, in which case they will stop along the sundry islands of the Mackenzie River. Snow geese will also move out prematurely if frightened by overflying airplanes. Once the geese have refueled in Alberta, they disperse by various routes toward their wintering grounds in California.

Each species has its own migration routes, its own food requirements, its own energy processes, and all the rest. Pectoral sandpipers that nest in soggy tundra meadows fly up the centre of the continent in spring and use coastal routes in fall. Golden plovers, after nesting on dry tundra, migrate all the way to the pampas of Patagonia. Each species is different; each has its special and unique needs. That is why it is not especially useful for us to attempt to generalize about the "birds of the arctic" for purposes of environmental impact assessment. No generalization is possible. There are for example several species of geese in the north, and although it is tempting to try to lump them and to speak of "the needs of geese", it can't be done. Each has totally different nesting and moulting areas, different food, different migration patterns, different social behaviour, and so on. Even different local populations of the same species will have different requirements and habits. With each new scrap of information we gather about living beings, the greater the revelation of our ignorance.

But the approach of winter is not the only hazard for migrant birds. Although food may be plentiful in the regions of their wintering habitat, many birds while fueling up for the spring journey northward may eat food contaminated by residual pesticides. These may be metabolized into the birds' systems as they draw on fat reserves during flight. Another ironic danger is that of arriving at their spring target too soon; their food supply must be thawed, activated, and waiting. To arrive too early is as dangerous as to tarry too late.

Swans, geese and ducks, and some other water birds, face additional hazards of their own in summer. Unlike most birds, in their annual moult (replacement of old worn feathers with new) they lose all of their flight feathers at once. Most birds shed and replace their flight feathers gradually, thus remaining airborne, but the waterfowl are grounded for a few weeks. Ducks and swans gather in flocks on the water, in sheltered bays, to await their replacement feathers. Snow geese remain in herds on the tundra. This is a physically demanding time, and obviously a vulnerable one. They can be herded and massacred like fish in a barrel, or even a disturbance can weaken them at a critical time before fall gathering. As well, some female waterfowl eat little or nothing while they are incubating their eggs, and depend on the subsequent flightless period after the young are hatched to feed and regain top condition. The growing of new feathers is an extra demand. Thus

41

there are pressure points throughout the season; there is really no "safe period".

Always there are hazards. There are arctic foxes, gulls and jaegers ever watchful for eggs or young birds. Once in a while a wandering wolf or wolverine might seize the opportunity to do some fueling of its own. Occasionally some trudging bear may find itself in a goose or duck colony; bears can consume a lot of eggs when they have a mind to. (Helpless to act, our field party once watched a barrenground grizzly do away with our entire supply of eggs—three dozen—in virtually no time at all. It also consumed frozen steaks and chops, also by the dozen, and bacon by the kilo.)

Always, weather is the key. Bears, foxes and avian raiders can all be coped with, but a late spring or an early autumn can be disastrous. Also there can be snaps of below-freezing temperature at any time during the summer. Yet the resilience of these beings is extraordinary. One amazing example occurred during a nesting survey of Lapland longspurs on the tussock tundra. Just at hatching time (about June 15) a nasty blow moved in from the Arctic Ocean. Fog, sleet, rain, wind, snow, and—for June—a horrendous wind-chill. Yet these virtually naked little beings, sheltered by the individual tufts of grass called tussocks in the lee of which each nest was placed, made it. Had they not, the parents *might* have had time to renest. Larger birds such as geese and swans, which have a comparatively long nesting cycle, do not have that opportunity. A year of bad summer weather can be a year of total breeding failure.

It seems that in the course of their evolution, arctic birds have been able to incorporate into their breeding cycles and population dynamics at least some capacity for compensation following occasional catastrophes. It is well known that in some years there can be a virtual wipe-out in some bird colonies, but after a few successive beneficient summers, and assuming no unusual mortality on migration or at the wintering grounds, populations can move back to "normal" (if there is such a thing) levels. Of course, the reproductive *potential* of plants and animals is often vastly greater than the number of offspring actually produced in a given year. The famous codfish with its millions of eggs is a prime example. Only a minute fraction of the eggs spawned ever grow into adult codfish, but the potential is there, held in abeyance, as it were, against emergencies.

A prime example of natural arctic "disasters" concerns the murres of Coburg Island in Lancaster Sound. Murres are seabirds, members of the auk family, that gather in stupendous numbers on a few favoured breeding cliffs. For food they depend completely on marine life in the waters below and near them. In 1978, David Nettleship, a seabird ecologist with the Canadian Wildlife Service, reported that a three-week delay in nesting brought about by adverse weather had caused colossal looses among young birds. And there could be no renesting. How often these natural blows to bird population occur, no one knows, but it would appear that as long as they do not happen too close together, the birds can handle them. In other words, murres have become adapted to occasional disasters. They cope.

Now, the critical question is how much *more* can arctic animals cope with? More, that is, than they already do. The murres offer a further case in point. In the fall of the year, their nesting finished, the murres of the eastern high arctic move away from their islands and channels east and southward, toward Labrador and Newfoundland. For decades, these offshore winter populations have been horribly decimated by oil discharged from ships. This is no accidental discharge; this is oil deliberately jettisoned by ships as a by-product of bunker-cleaning. Also, between 1968 and 1973, from 500,000 to 750,000 murres were killed in Danish fish nets off Greenland *every year*. The fact that murres would rather swim than fly makes them especially vulnerable. The Inuit of Greenland take about 75,000 each winter for food. If there should be a breeding "bust" caused by the weather as in 1978, together with an especially heavy winter-kill of adults, no one can say how long it might take for a murre colony to rebuild—if it ever could. We do know that the birds seem to be able to cope with periodic losses of young; losses of adult breeders is another thing altogether. How many additional unnatural stresses the birds might be able to accommodate we shall probably never know, except by hindsight.

There are other such examples. Eider ducks, which nest throughout the major portion of the arctic, spend the winter in both Atlantic and Pacific waters. On their return to their nesting areas in spring, great flocks of birds arrive over the Arctic Ocean immediately after (it seems almost simultaneous) the first long cracks or "leads" of open water occur in the sea ice. They settle down on these narrow stretches of water to rest, and presumably to feed. Occasionally there will be an unusually late spring, and the birds may arrive before the open water has appeared, or after some storm has temporarily closed it off. This has been known to be fatal to great numbers of birds. One year Thomas Barry observed no less than 100,000 eiders—10 per cent of the regional population—die on the Beaufort Sea of starvation and exposure in this manner. But as we know, in natural circumstances the population does seem to recover. One cannot help speculating on the outcome, however, were some oil spill trapped under the Beaufort ice to surface in spring in the open cracks, to which the moving birds are unswervingly attracted. Such speculation does not seem out of place in the 1980s, with both increased drilling and tanker traffic in the cards for the area.

THE MAMMALS

The occasional misfortunes of the geese, the murres and the eiders—which are merely highly visible manifestations of a process that is going on continually—seem to be echoed, but on a slightly more regular basis, among small mammals. (Large mammals are subject to periodic population "busts" also, but we do not know very much about them.)

You have probably heard, both in fact and fiction, of the wild and spectacular changes that occur in populations of the Scandinavian lemming, a small mouselike creature whose mass emigrations at the height of a population buildup have given rise to some of the most fanciful writing (and even film-making) of our time. The brown lemming of the Canadian arctic is quite closely related to the Scandinavian species, and the numbers of both brown and collared lemmings fluctuate on a schedule of approximately every two to five years. In *The Mammals of Canada*, Frank Banfield describes how the population of the brown lemming "slowly builds up over a period of two to four years provided climatic, vegetative and

44

intrinsic factors are favourable. Eventually the high breeding potential produces an excess population, which may overutilize the food supply and force the lemmings to emigrate in order to survive. . . . The emigration may carry lemmings into town-sites, across lakes and bays, and out over the frozen sea, where their carcasses have been observed more than ten miles from land." The population then "crashes", and starts to rebuild.

Such oscillations are not necessarily "across the board" in any one year. Lemmings may be scarce to the point of apparently total absence in one region, and still reasonably visible in another. But where the lemmings are at the bottom of a crash there will be concomitant reductions in the numbers of the animals that prey on lemmings—snowy owls, for example, or rough-legged hawks. Also affected will be the pomarine jaeger, a rakish piratical seafowl that looks like a cross between a hawk and a gull and whose fortunes are intimately dependent upon those of the lemmings. When the latter are increasing most profusely, there will be an unusually high success rate in the reproduction of those animals that subsist on lemmings. A rich food base means a higher proportion of survivors among the young predators. They increase thus just a step or two after the lemmings. When the lemmings eventually collapse, the birds may overshoot the mark somewhat, and there will be many hungry mouths unsatisfied. It is in such times that birdwatchers encounter unusually large numbers of snowy owls, for example, in southern Canada. Their breeding success was such that it ultimately outstripped a diminishing food supply.

But there may be even more to the repercussions of lemming oscillations than that. Consider how a local superabundance of lemmings would begin to have an impact on the health of the grassy tundra vegetation that supports them. This effect could be transmitted to other species that might depend, in one way or another, on the same vegetative cover. Also, when lemmings temporarily vanish, species that normally depend on them such as hawks and owls, foxes, and jaegers, may be forced to turn to alternative foods, thus exerting pressure that is heavier than usual on those other species. So it is that both the presence of the lemming in high numbers and its total absence can be transmitted through the food web in different ways. Such a process could prolong the effect of the lemmings' absence, for example, over a longer period of time than we might expect. Other species could still be suffering, even when the lemmings are already rebuilding.

As species, lemmings cope with the arctic conditions by having an extraordinarily potent reproductive potential (which ironically enough is also the seed of their collapses). As individuals, lemmings also have special adaptations to northern peculiarities. Both arctic species of lemmings have networks or labyrinths of tunnels in the vegetation during the summer; in winter their mazes are beneath the snow. (Lemmings do not hibernate; they are active all year.) The collared lemming, like the varying hare or snowshoe rabbit of the boreal forest, turns white in the winter. Even more remarkable than this (lemming-eating weasels also change the colour of their coats) are the strange "snow shovels" that develop on its claws for wintertime use. The two centre claws on the forefeet develop little paddle-like structures which greatly lengthen and broaden them, no doubt aiding the lemming in snow-tunneling in much the same way that the winter "snowshoes" of the ruffed grouse of the south help it to get about on a soft surface.

From snow shovels it is a short step to snow scrapers, as epitomized by the great forehoofs of the muskox, which it uses to get at its winter browse. It doesn't have to depend on them entirely, however, for in winter the muskox herds will seek out the most windswept places, where the need to scrape snow is least. But if you are going to seek out hills and open country in the middle of an arctic winter, it pays to be prepared. The muskox is superbly insulated, thanks to a dense, deep undercoat as soft and fine as cashmere, which is overlain by a long shaggy robe of coarse stringy hair. The long overcoat has been described by everyone who has ever seen a muskox as swinging like a Scotsman's kilt and sporran.

As with the birds, the insulation of the muskox serves more to keep heat in than to keep cold out. And if the retention of body heat is the prime objective, then it pays not to have too many loose appendages. Seen at any distance, your first muskox reminds you of a huge black shredded wheat. Its shaggy shape is boxy and massive, with little apparent relief. At closer inspection there are of course the oddly shaped horns and the suggestion of a nose, but in general the impression is of square solidity. A muskox might just as well be

46

without ears and a tail, for all you can see of them. Protruding extremities lose heat, and can freeze. The principle involved here is quite simple. The boxier an animal is, the greater is the ratio of internal volume to surface area. A long-legged, long-necked, skinny animal like a giraffe has a relatively small body volume in relation to surface area, and loses heat readily—as a giraffe, having in mind its environment, properly needs to do. The muskox needs to hold all the heat it possibly can, and by being rather high in volume by comparison with its external surface area, does just that.

Indeed many arctic species or subspecies tend to be built along somewhat heavier and bulkier lines than their more southern relatives. Barrenground grizzlies look a bit more rolypoly than those of the south. Peary's caribou of the high arctic islands are a good deal stockier than the mainland forms. Compare a short, sturdy Inuit and a tall, slender Masai of East Africa for that matter. They are very different in shape, and for excellent reasons.

(The late Lester Snyder of the Royal Ontario Museum worked out how this principle applies to birds. As we have seen, the northern regions are characterized by loons, auks, waterfowl, birds of prey and so forth. Snyder estimated that a good 60 per cent of the bird species north of the treeline are at least of the bulk of the common pigeon. By contrast, only 30 per cent of the birds of Ontario are of that size. This general rule appears to apply to cold-blooded animals as well as to birds and mammals.)

Although it might sorely try the imagination, put yourself in the position of a muskox during a howling winter blizzard on Devon Island. Granted, you have the right build for it, and you are equipped with the most marvelous protective windbreaker and underwear ever designed. Even so, survival must require more than that. In a monograph prepared for the federal government John Tener calls attention to "the customary slow and deliberate movements of muskoxen" as an additional adaptation. Unnecessary exertion would waste fuel. A further advantage is gained by the fact that the muskox droppings in the winter are "in the form of dry, round pellets, in contrast with the loose, soft material of summer. This is partly the effect of the much drier winter foods, and possibly of a low water (snow) intake, but also conserves the energy which the animals must use in winter to convert snow into water."

47

Such evolutionary adaptiveness seems nothing short of miraculous, and perhaps it is. But there is always a price to be paid for specialization. Stolid and slow moving as they are, muskoxen have no way of escaping from swift tundra wolves. Their formidable horns are excellent defence from the front, of course, but no muskox can protect his rear. So we have the famous circle into which the muskox herd forms, all horns presented outward, not unlike the embattled wagon train of the cowboy-and-Indian films, and for many of the same reasons. A wolf has slim chance of penetrating that formidable ring, and unless he and his companions can break a young calf out into the open, more often than not the defensive manoeuvre works. Wolves will of course pull down solitary animals, usually old bulls wandering about singly or in pairs. Of course the defensive circle of the muskoxen makes the animals openly and pathetically vulnerable to gunfire, and standing massacres have been the sorry story of the species since firearms first appeared in the north.

In contrast to the peaks and hollows of lemming fortunes, which are well known and to some extent understood, those of the muskoxen are not so well documented. Over long periods of years there have been inexplicable plunges in the numbers of muskoxen and changes in their geographic distribution, virtually without warning. They have disappeared entirely from some areas and reappeared in others. We know of the effect of historic mass slaughters by both natives and whites in various places, but this cannot explain why the animals suddenly vanish at times from apparently favourable places in the arctic islands. Nor can it explain the occasional discovery of their remains lying about with no apparent cause of death. No doubt disease is a factor (but that is not much help, because disease is always a factor in animal populations). There is obviously a strict limit to the number of these animals that can be supported by their food supply, and always—as with the birds and all the rest—there is the unpredictability of arctic weather. Clearly no living being that is so finely tuned to such a marginal existence can tolerate the least additional pressure, from whatever direction it might come. Probably it has always been thus, and the arrival of men with firearms, in many places at least, simply tipped an already precarious balance between survival and its opposite. No animal on the face of the earth is "adapted" to us.

The muskox is vastly outnumbered in the arctic, and probably always was, by the caribou, the world's northernmost deer. There are a number of recognized races or subspecies, of which three are found in the arctic. The famous barrenground caribou of the vast continental interior, Baffin Island and south Greenland is viewed as a different race from the herd in the northern Yukon and Alaska, while the shorter, stockier Peary's caribou occupies the high arctic islands. Although the barrenground population is the best known popularly, in recent times by far the most scientific attention has been given to the Yukon herds, especially the famous Porcupine animals, which were the focus of much deliberation for the Berger Commission in the course of the Mackenzie Valley Pipeline Inquiry in the mid-seventies.

Caribou are unusual among deer in that both sexes have antlers, although those of the males are bigger. During their annual growth, deer antlers are encased in "velvet", a skin-covered network of blood vessels which after the antlers have been nourished to full growth, dries, withers, tatters, and falls away. One might guess that the antlers of females are smaller at least partly because females are pregnant during the period of antler growth, and at that time the females have more pressing nutritional requirements. One could also guess that the antlers of the pale far northern race, the Peary's caribou, are much less splendid than those of the mainland animals simply because excessive loss of blood heat through the growing velvet would not be adaptive. Their antlers are shorter, straighter, and more slender; many females carry mere spikes, or do not have any antlers at all.

Wolves are the natural predators of caribou, wherever these deer are found, from the high arctic islands to the continental barrens. A few newborn caribou may be taken by wolverines or bears, or the occasional lynx. There is an ancient and honourable relationship between caribou and wolves, as there was at one olden time between caribou and men. The dog and the deer have evolved together; the awesome complexities of their behaviour and their life cycles have become adjusted over evolutionary time to the essential presence of each for the other. A given stretch of tundra is "designed" to sustain (or, if you are of an economic bent, "capable" of sustaining) a certain number of caribou over a certain number of months. We must remember, however, that the capability of tundra to sustain life necessarily includes the lives of wolves, and of the eagles and ravens and foxes who follow them, and of the invertebrates who follow *them*, and of the decomposers who feed whatever morsels of nourishment are left back into the hungry soil for the support of next spring's flowery tribute to the process. All is of one piece; wolf and caribou couldn't do without each other.

In *Of Wolves and Men*, Barry Lopez tells of the ancient and honourable relationship of which we speak here. When the relationship comes to the ultimate encounter between individual wolf and individual caribou, there is a timeless moment he calls the "conversation of death". In that moment there is an "exchange in which the animals appear to lock eyes and make a decision. . . . It is a ceremonial exchange, the flesh of the hunted in exchange for respect for its spirit. In this way both animals, not the predator alone, choose for the encounter to end in death. There is, at least, a sacred order in this. There is nobility. And it is something that happens only between the wolf and his major prey species. It produces, for the wolf, sacred meat."

The prey animal that has been singled out by the predator knows that it has been singled out. Indeed it has participated in the identification of itself as prey. There passes between pursuer and pursued a communication which we can never understand, and that most of us would prefer not to acknowledge. But it happens. The communication is the acknowledgement of what will now take place. On the other hand, you can see caribou grazing quite nonchalantly as a wolf trots by, secure in the knowledge that the moment is not now. You can observe exactly the same relationship between the gazelles and other antelopes of Africa and their predators—lion, cheetah, and so on. It is something very old and of supreme perfection that "civilized" men, the predator-hatred deep in the bowels of their culture, do not wish to see, much less try to understand.

Like all other arctic beings, caribou have developed physical adaptations to local conditions. As demonstrated by the remarkable variety of uses to which they were put by Inuit and Indians, the pelts of northern deer are extraordinarily efficient. Each individual hair is hollow, filled with air. This not only provides extra insulation but also conserves the energy of the caribou by providing an unusual buoyancy in swimming, a great deal of which has to be done in the course of migration. A further aid in swimming is provided by the animal's very large (for a deer) hooves, often wider than they are long, and strongly curved, which give a pigeon-toed effect. There are very large dew-claws for extra purchase on soft ground, and the sharp edges which develop on the hooves for winter result in "'non-skid' support", as biologist John Kelsall calls it, on the ice. Thus caribou are equally proficient in navigating ice, snow, mushy ground, mud, or water. In the course of their migrations they encounter all of these, as well as clouds of flies and other insects, occasional predators, and frighteningly variable weather.

Caribou are almost perpetually on the move, in more or less predictable directions between their wintering and calving grounds, with many side forays and other excursions complicating the routes. It is an interesting experience to know perfectly well that there are thousands of moving caribou in your area, yet not to be able to find them. You may have seen them from the air the day before, but that can be of little help today. The tundra is big and rolling, with many ridges and valleys, and the caribou don't confide in us their selection for any particular day. In their constant movement they are very much like, though not nearly so conspicuous as the African wildebeest, whose trails over the short-grass savannah remind you immediately of caribou paths on the dry tundra. This unceasing travel is very important. Great numbers of animals, if they stayed in one place, could have an unfortunate impact on their limited food base. So they keep moving, and the effect in one spot is never so great that the plants cannot recover.

The mainland caribou move out onto the tundra in spring and early summer to drop their calves. In the fall they return to the relatively more benign surroundings of the northern edge of the spruce forest. In both habitats they depend overwhelmingly on one food source. As Frank Banfield says, in *Mammals of Canada*, "If horses are called 'hay-burners', reindeer should be called 'lichen-burners', because these lowly plants are the mainstay of their diet, especially in winter." The caribou lichens (often misnamed "mosses") are slow growers, and the impact of the grazers must be evenly spread, not only over space but also over time. Thus the herds don't return to the same wintering areas every year, or take exactly the same migration routes every year. It gives the lichens a chance to recover.

Caribou lichens are seriously vulnerable to fire. This is not (or has not yet been) a significant problem on the tundra, where a mosaic of dry areas tends to be interspersed with wet marshy spots. But in the forested wintering grounds fire can be disastrous. This danger becomes more and more severe as greater numbers of people appear in the spruce zone, not only increasing the fire hazard but also reducing the number of optional wintering areas available to the herds. Various estimates have been reported by John Kelsall and others of the length of time required for suitable lichen growth to return after fire; these range upward from fifty to one hundred years or even more. The gradual increase of fires in the north and the resulting decrease in available winter habitat will no doubt prevent the barrenground caribou from ever approaching their original population of some two and one-half million, even if other agents of destruction, such as shooting, were to cease altogether.

The caribou of the high arctic perforce do not migrate to the same extent as the mainland herds. They have no spruce bush to resort to, and the only cover or protection from wind-chill is that offered by the contours of the land. They do move, seasonally, to dependable food sources. We should not conclude, however, that they are necessarily stuck on their islands for keeps. At some time or another in the past at least small numbers must have travelled from island to island over the winter ice, perhaps occasionally swimming in salt water, in order for them to have achieved the distribution we now observe. Presumably barrenground caribou made it by some ice bridge to Baffin Island from the Melville Peninsula at some unknown (but perhaps more recent) time. Perhaps there has even been movement between Greenland and Ellesmere Island at some time, although caribou are thought to have arrived in the Americas originally by the Bering land-bridge to Alaska.

The archetypal caribou migration is that of the Porcupine herd in the northern Yukon. After having spent the winter in a surprising variety of habitats—all of them relatively sheltered—the herd moves from areas south and west of the Mackenzie Delta out onto the coastal plain for calving. The calves are born as the herd moves, from late May to the middle of June. Within a day, young newborns can keep up with their elders, fording icy rivers with the best of them. Gradually the great herd (split into subherds), moves north-west along the North Slope toward Alaska. Insects annoy the animals frightfully, and many will move right to the seacoast to take advantage of an almost constant wind. It is a strange experience to see caribou walking along bars of mud or gravel with their great feet sloshing through salt water. (A photo was taken one day of a bird researcher sitting on the beach, steadfastly counting ducks through his binoculars, with a female caribou curiously nuzzling the over-long curls at the back of his neck!) Later, as the herds move back toward the foothills of the Alaskan mountains, the caribou will cope with insects by traversing windswept ridges wherever possible.

COPING ON THE YUKON COAST

West of the Mackenzie Delta, a carpet of rich tundra extends ribbonlike along the coast, roughly parallel to the Richardson, British and Brooks series of mountain ranges inland from the Beaufort Sea. This is the area where the snow geese gather in fall, after the caribou have moved off on the last loop of their annual trek. A series of northerly flowing rivers drains downward from the mountains across the tundra to the ocean. Time, including spring river torrents, storm wave action, ice grinding and imponderable geologic events, has created along this coast a bewildering complex of braided gravel deltas, islands, bars, beaches, spits and lagoons adjacent to the usual tundra lakes, ponds and sedge marshes. It is unimaginably rich in life—a slim and attenuated oasis that is unique in Canada and (apart from adjacent Alaska) in the world.

In season, polar bears stalk the beaches and grizzlies forage in the inland tundra. Yellow-billed and black-throated loons fish along the edge of glittering ice. At least once, a soaring sea eagle cast its moving shadow over waving cottongrass and yellow dancing arctic poppies. There is an ancient fox warren at Nunaluk Spit, where the flowers bloom luxuriantly not only in the foxes' garden but also around an abandoned Inuit whaling camp of long ago (Nunaluk lagoon is strewn with the great vertebrae of bygone bowheads). Snowy owls nest at Komakuk Beach, and great flocks of sea ducks known as scoters spend their flightless moult in the sheltered bays of Herschel Island, where not long ago a brave Inuit man called Jonas hunted walrus alone, with a harpoon.

Along the gravel beaches and spits there nest glaucous and Sabine's gulls, common eiders, and arctic terns. Snow buntings and ptarmigan frequent the wider beaches. In the wet sedgy tundra are nesting semipalmated and pectoral sandpipers, phalaropes and geese. In the wet tussocks and among the frost polygons, are whimbrels and dowitchers. On the drier tundra, in the flatter areas golden plovers are nesting, and the rare buff-breasted sandpiper (a southern birdwatcher's bonanza) holds its communal courtship performance on the gentle slopes where the best-drained tussocks cluster. Marsh hawks quarter the open plain, and swans call from the tundra lakes.

The flight of snow geese in their pre-migration feeding and gathering known as "staging", though unforgettable as a spectacle, is merely one aspect of a much larger migration. The coastal complex is used by literally millions of birds both as a corridor and as a stopover: loons, swans, geese, ducks, hawks, eagles, falcons, cranes, plovers, sandpipers, phalaropes, jaegers, gulls, guillemots, owls, ravens, and assorted songbirds.

The shoreline lagoons of the Yukon coast, sheltered from the Beaufort by the barrier beaches, are lower in salt than the sea itself. They are made brackish by spring run-off, and especially by the great flows of snow-melt from the mountains. The result is a peculiar mix of waters, and thus of nourishment upon which birds and other animals depend at the various seasons. There is an extraordinary (but not at all well-understood) production of invertebrates and fishes in these lagoons, which are used as foraging areas by loons that may nest on small lakes several miles inland. There is exquisite timing in the way these areas are used sequentially by different species over the year. One example is the little Baird's sandpiper, which after nesting along this coast and through the high arctic islands, appears after its young are flying to forage in fast-disappearing tiny pools in coastal deltas. One can assume that the invertebrates they require at that season appear at just the appropriate moment, thanks to the right level of drainage at the right time.

In the fall, throngs of northern phalaropes feed skipping prettily along the seaward side of the beaches, always with attendant jaegers, which frequently chase and catch them on the wing. One phalarope thus pursued took refuge under a card table set up on the beach, at which a researcher was working. It settled close to his feet while a frustrated jaeger circled, but not too close, voicing its displeasure. Another phalarope, dodging another jaeger, flew to the ground between two of us who were pushing our float plane off the shore into the lagoon. There it stayed until the jaeger set off in pursuit of yet another phalarope.

Bearded, harbour and ringed seals ride the ice along the coast in spring and summer, and the occasional Pacific walrus may still wander into Canadian waters from Alaska. Pods of orcas ("killer" whales) sport dolphinlike offshore from time to time, and there is a pitiable remnant of the once numerous bowhead, a whale that was hunted here to the point of "economic extinction" by about 1900. It gives me pleasure to relate, however, that there are still many white whales along this coast, especially in summer, when some five thousand of them gather from the Beaufort Sea near the western side of the Mackenzie Delta to have their calves. Sometimes they penetrate deeply into the labyrinth of channels and lakes of the Delta region. On the east side of the Delta, in the innermost reaches of Liverpool Bay south of Tuktoyaktuk, white whales have been known to linger too long in the inshore passages and openings, and to have been trapped by an unusually fast freeze-up. The breath of life denied to them by the grimly closing ice, the whales eventually drowned. Such are the exigencies involved in coping, or failing to. The whales "gambled" that they could get out of there before the freeze, and they lost.

COPING IN LANCASTER SOUND

White whales have a very wide distribution in Canada, from the highest open water in the arctic, south to the St. Lawrence Gulf and the River itself, which they penetrate as far as Quebec City and even beyond. These southern populations represent a distinct group that does not overlap with the more northern animals. Most whales do not of course fall victim to the misadventure that has overtaken those of Liverpool Bay on occasion, and they spend the winter wherever there is open water. In summer the white whales of the eastern arctic enter the opening waters of Hudson Bay, Foxe Basin, and especially Lancaster Sound, through which a good third of all the 30,000 white whales in North America pass each year, on the way to traditional calving and feeding areas.

Narwhals, the unicorns of the sea, pass through here too. The long spiral ivory tusk of the male narwhal (a modified maxillary tooth) has been admired and coveted by men for centuries—with the usual consequences. Not much is known about the narwhal, including whatever the reason may be for its attraction to ships. Like several species of their distant dolphin relatives, narwhals will approach and accompany a ship. This can get them into trouble, as it did one silent arctic evening as a large cruise ship entered the harbour at Pond Inlet on Baffin Island. As the ship very quietly came to berth, a sudden fusillade of cracking gunfire from hunters on the shore ripped the silence. A half dozen or so narwhals had been keeping us company.

As a biological oasis in the vastness of the Canadian north, Lancaster Sound is the marine equivalent of the Yukon North Slope. It seethes with life above and below the water's surface. The eastern opening of the "Northwest Passage", the Sound is bounded on the north by Devon Island, on the south by Bylot Island and the north coast of Baffin Island. Flowing past immense looming cliffs and mountainsides, sheer rock-faces and jagged promontories, its often formidable waters enter on the west between Cornwallis Island and Somerset Island, and empty into Baffin Bay on the east.

Here, for reasons which have not really begun to be explained, there is a production and concentration of living material that is unparalleled in high arctic seas. No doubt this bursting profusion of marine life is related, as it is in other parts of the world, to upwellings and mixings and foldings of waters moving in different directions, at different temperatures and depths, at different seasons. Nutrients may well arrive here from every point of the compass, and their convergence would result in their "recombination" and profuse reincarnation because of the intrinsic characteristics of a very specific local situation. But no one yet knows how, or why, or—most especially—why *here*. The eminent marine biologist Max Dunbar has remarked that the question "Why should it be so productive?" was not even addressed, much less solved, by a recent "environmental impact study" conducted in Lancaster Sound. You might well ask what such an apparent oversight has to tell us about the "environmental assessment" process. That will be discussed in Part Two.

As in all seas, the basis of the life structure of Lancaster Sound is the aggregation of microscopic plants and animals that collectively are called plankton. Phytoplankton are plants; zooplankton are animals—although sometimes the distinction becomes blurry. Since their quantity is so great, normal arithmetic expression becomes meaningless, and plankton are usually assessed in terms of numbers per volume unit of water, or their weight per unit. In their numberless upwelling hordes, they sustain each other and the larger invertebrates up the line, culminating in the fishes, birds and mammals of the Sound. On the ocean floor there are starfishes, sea cucumbers, urchins, and corals, above which weave gigantic Greenland sharks. On the underside of the ice there are special kinds of algae for whom this is home and habitat; they are grazed upon by all manner of invertebrate animals. Like so many algae, these minute plants "bloom" in startling quantity, in their season, and form a basic and fundamental part of the food webs of these waters. Then, there are the small fry of the arctic codfish—which, according to Allen Milne and Brian Smiley, "may be singly the most important link between marine invertebrates and higher birds and mammals." (What that critical link may be along the Yukon coast no one yet knows.)

The codfish are able to make direct use of virtually the entire spectrum of marine food sources—all the way from phytoplankton in the springtime to shrimps and other crustaceans, a long list of miscellaneous other invertebrates, fish eggs and fry, and of course lesser codfish. The small cod themselves have been described as the "primary" food source for such seabirds as murres, fulmars, and kittiwakes. Larger cod support ringed seals, and ringed seals are in turn the staple food of polar bears—and, at least in times past—the Inuit. Just as everywhere else, all things are of a piece.

A very important basic group of invertebrates—the molluscs—provides for eiders and scoters as well as walruses and bearded seals, both bottom-foragers. When you think that a bearded seal may have a mass of over 350 kilograms and a walrus over 900, the richness and the continuity of clam abundance are obviously essential. But no walrus, seal, whale or other mammal can survive, even on this bountiful supply of food, for longer than a few weeks each year. Autumn freeze-up is never far away. Just as it is everywhere in the north, the key to everything else is timing; you arrive on time, and you leave on time—or else.

We have seen the havoc that a late spring can play with breeding seabirds such as murres. But losses of young birds can occasionally be tolerated, as we have discussed, and over the longer period most species seem to be able to cope with natural setbacks. Certainly there is an overwhelming abundance—a teeming—of seabirds in Lancaster Sound during the ephemeral summer. Indeed the numbers are so great that estimates of populations tend to vary with the technique used, the observer, and the year. But here are some examples: at Cape Hay, on the north side of Bylot Island at the eastern entrance to Lancaster Sound, some half a million murres (twenty years ago there were a quarter million more than that); another half million on Coburg Island; on the staggering cliffs of Prince Leopold Island, a mere 140 thousand. In Lancaster Sound and Barrow Strait at its west there are about a quarter million fulmars, the only representatives of the "tube-nose" (albatross) family in the Canadian arctic.

In the economy of the human marketplace, abundance is equated with cheapness, and there is a premium on scarcity. The sheer number of birds in Lancaster Sound might seem to make them commonplace, expendable. But the overwhelming majority of them is confined to a handful of very particular sites. Biologically, each of these precious few concentration points is beyond price, because there aren't any more of them.

Essentially, everything that happens in a biological sense in Lancaster Sound depends ultimately on the behaviour of the ice in a given year. The massive ice barriers must fracture in May and June so that white whales, narwhals, seals and walruses, and feeding seabirds may have access to its teeming productivity at the right times and in the right amounts. As on the Yukon North Slope, any "slippage" in the regime of freeze and thaw can have dire results, at least for one year. These disruptions do occur, how often we don't know, but local communities are geared to cope with them.

The single species that at least on the surface of things shouldn't have to worry about the presence of ice is the polar bear, who is supremely well-adapted to traverse it. But even the world's greatest carnivore has to find the end of the ice somewhere; there has to be an edge, and at least some open water, because that is where the seals will be. Bears travel constantly over land and ice, and can swim long, cold miles also. They have to stay in touch with the seals, or they will perish.

Those female polar bears that are pregnant do not brave the ice all winter; they hole up somewhere on the mainland or even on the ice itself in some snow-covered hollow to have their cubs. By spring the young bears will be large enough to keep up with their mothers when travel has to be resumed. Male bears, on the other hand, are usually on the go all winter long, their swinging, rolling stride taking them extraordinary distances over the uneven, blizzard-torn ice. By spring, some of these large males are very hungry; if you spot a lanky, long-legged, slab-sided bear at that time, give him plenty of room.

Numerous birdwatchers visiting the north have been indebted to the polar bear for "producing" for their lists one of the loveliest birds that flies—the elegant and graceful ivory gull. Small flocks of these stunning birds often attend old Nanuk on his wanderings, feeding on discarded seal offal and other droppings from his table. Indeed it is said that in winter, if the polar bear does not make a kill soon enough to satisfy the urgent, gnawing metabolism of the birds, they may well subsist, at least for a time, on droppings not from the bear's table but from the bear himself. That is another way of coping in times of stress, and it works.

THE PEOPLE

In the past, the Inuit hunted bears for food (both for themselves and for their dogs) and for their enormous skins. Bear hunting is still an important source of revenue for the Inuit of Resolute Bay on Cornwallis Island and of Pond Inlet at the northeastern corner of Baffin Island. There apparently remains a reasonably good population of polar bears in the Lancaster Sound area (you can sometimes see four or five at a time at the eastern corner of Devon Island) as evidenced by the rapidity with which commercial kill quotas are filled annually.

There has been much nostalgic and even romantic writing (mostly by Anglo-Saxons) about the bygone days of the Inuit and of their remarkable ways of coping with arctic environmental circumstances before the arrival of the white man. We do not know how long humankind has lived in the far north, except that the years may number in the thousands. Adaptations to extreme environments, as we have seen, must involve anatomy, physiology, and behaviour. Northern peoples have been shown to have developed, over time, anatomical and physiological differences from more southern people which gave them a greater physical survival advantage than that enjoyed by intruders from the south. But that was only part of the story.

Much more important was the evolution of a "culture" (including social behaviour, customs, traditions, beliefs, technology, and all the rest of it) that was appropriate to the environmental conditions in which the people lived. Wherever they are, human beings "cope" at least as much by way of culture as they do by way of physical adaptation. Indeed technology (the "how-to" of clothing, weapons, tools, food procurement and preparation, shelter, transportation, etc.) is brought to bear as a substitute for those physical attributes which *Homo sapiens* the species simply cannot have in sufficient variety to meet the requirements of the infinite number of physical environments in which he has found himself. Most important of all, perhaps, are traditional ways not only of doing things, but also of *perceiving* things such as the external biophysical environment, the social environment, and perhaps even oneself.

From the biological point of view, environmental *appropriateness* is the fundamental prerequisite for the continuing survival of a species. But appropriateness, like the notion of ecological "niche", is usually inferred after the fact, not predicted. The inappropriate species (the rabbit in Australia, for example) is visibly and obviously identifiable as inappropriate only after it has wreaked its environmental devastation. Another form of inappropriateness is that which we ascribe to the dinosaur on the incontrovertible evidence that he is not present in today's environment. On the other hand, the positive side of all this—appropriateness—is concluded when we see a continuity of apparently well-adjusted relationships between a species and its environments. The muskox and the murre and the white whale "fit" the arctic. Presumably, so it was also with the Inuit, once upon a time.

It seems to be agonizingly difficult for contemporary, "civilized" mankind to accept the implication of "environmental appropriateness". The lemming "fits" only by virtue of its periodic population collapse, which is offset by its inherent ability to recover over a relatively short time. Lemming appropriateness is achieved by way of very considerable individual "sacrifice". Were it not for population crashes, the animals would destroy their own food base, and there would be no more lemmings *at all*. Other innocent bystanders, deprived of their lemmings, would vanish also. But that scenario is not what life is all about. Instead of disappearing completely, the lemmings always have a small reservoir, from which the rebuilding and reconstruction of lemmingkind can proceed. Many individuals are expendable if the ultimate goal, as we have to assume it to be, is the perpetuation of the lemming.

The screaming, honking, growling bazaars of murres on their towering sea-cliffs in the mist, are able to maintain their own continuity because their genes have learned how to cope with periodic losses of young nestlings. They may not, however, know

61

how to deal with significant losses of breeding adults. The eiders of the Beaufort can work their way out of huge losses of adults on the frozen sea, so long as their nesting colonies are able to resume activity with the help of one or two favourable summers. But many individuals have been lost along the way. Many individual caribou each year conclude their sacred pact with the wolves, but the integrity of the population and thus of the species is maintained.

So too it must have been with the Inuit in the "olden days". Life was hard and unrelenting, unquestionably, and as with any other arctic beings, mistakes were costly and conclusive. But over the passage of time there developed a "balance"—no doubt at times an uneasy one—between the people and the environment that sustained them. There were not many people, of course, because the far north cannot sustain many people. We can conclude with confidence that the traditions and the technology and all of the other ways of the Inuit, before the white invasion, were appropriate to the environmental circumstances. Otherwise the people would not have been there. We can speculate that there were, over numbers of years which we cannot know, for the Inuit as for all other arctic beings occasional collapses or "busts", whether caused by weather or by disease or by that uniquely human specialty, war. It would be difficult to imagine that before the coming of the white man the history of the Inuit did not have its periodic catastrophes. But the Inuit survived. They were appropriate. We know that *because their environment survived also.*

The human species is of course the ultimate consumer. When some technical innovation gives a people a new and unprecedented advantage over the species they consume, the consequences can be spectacular. All over the world, the introduction of firearms to the hands of human beings of all races had immediate and terrible results. The end of the great whales came with the explosive harpoon and the factory ship. All over world too, the introduction of currency (the perception of the animal not as sacred meat but as cash) has had parallel results. The "conversation of death" has nothing to do with money. White whalers and traders (and also missionaries, to whom we will return in Part Three) brought to the highest latitudes not only exotic diseases with which the original people could not cope, but also exotic techniques with which the people's environmental base could not cope. That there was also delivered an ideological payload with which Inuit traditions, culture and perceptions could not cope will be shown in a more appropriate place.

Because European intrusion came when it did, we shall never know whether native human beings and native non-human beings had ever reached an equilibrium in the north. We cannot know, now, what lasting effect the human presence may have achieved long before the arrival of the Europeans. Perhaps man had already eliminated species or (more likely) local populations, as the evidence shows he did in other parts of the world, long before the European influence. Perhaps not. All that can be said with any degree of confidence is that northern people had for generations demonstrated their ability, through cultural adaptation, to survive in the face of what still seem to the European mind impossible conditions.

THE "FRAGILE" NORTH

But environmental conditions in the north are not impossible, as all the evidence shows. There are changing fortunes, to be sure, oscillations and swings and yaws and booms and busts both over the short term and the long, depending on the species involved. When we reflect upon it, it seems that the much-alleged "fragility" of the northern environment is not so much related to the numbers of species of plants and animals, or to the populations of those species, or to any inherent flaw in the life process itself. If we must use the term "fragile" we should understand that we are talking about an environment of *extremes*, not of weakness. The one characteristic that is deeply inherent and ineradicable is its *unpredictability*.

Arctic communities are uncritically labelled with the pejorative "unstable" simply because they oscillate. Even worse is the statement that the arctic is "simple" because it does not have as many moving parts in the form of species as a tropical jungle has. We also hear much about the "low carrying capacity" of the north. What researchers or the press really mean when they say that is that the arctic cannot tolerate very many *people*, which is true.

Such charges are commonly levelled at the north, one might speculate, because the north cannot be accommodated within much of the conventional wisdom of ecology. It is useful to remind ourselves that most of such wisdom arose neither in polar nor equatorial regions, but in temperate regions, which are the most resilient there are and can absorb more human insults and other interventions than any other communities on the planet. It is useful to remind ourselves also that ecology does not cast up quite so many generalizations as the popular press might lead us to believe.

Few pop-ecology "buzzwords" have become so pervasive as "stability". Let us consider "stability" for a moment. Picture a high-wire performer with his long pole, traversing the abyss. A "stable" high-wire performer is a dead one. His secret is not his strength or his speed, but his feather-light flexibility, his teetering, anticipatory ability to offset momentary jiggles, his instantaneous adaptiveness.

Let us think too about "simplicity", another pejorative. Because there are fewer individual players (species) in the far north than there are in some of the more crowded arenas of the world, we see the story that is told as somehow less complex, less sophisticated, less subtle, less interesting, less "developed". At this point one might compare a Mozart quartet with the Boston Pops playing John Philip Sousa. The fact is of course that the complexity of a single cell (much less that of a termite colony or an earthworm) is far beyond our understanding. Yet we persist in seeing "simplicity" in the arctic, and "low carrying capacity"—the latter directly in the face of the unmeasurable productivity of arctic seas.

It would seem that much of the alleged "fragility" of far northern regions results from our use of inappropriate criteria. The ecological pronouncements that arise from quite different and thus incomparable regions of the world simply do not apply where they did not originate. This is not to say that the far north is not vulnerable to disturbance. It is—perilously so. But the reasons are not those that apply in other places, and the arctic will be doubly vulnerable so long as the assessors of environmental impact continue to apply criteria that do not fit.

12 *The boxy, massive muskox is supremely adapted to its life on Ellesmere Island.*

13

14

5

13 *Oil drums on Ellesmere Island.*

14 *Dome Petroleum's subsidiary,* CANMAR, *uses drillships in the Beaufort.*

15 *Inuit at school in Inuvik.*

16 *(overleaf) The Richardson Mountains south and west of the Mackenzie Delta.*

17 *Male rock ptarmigan, Bathurst Island.*

18 *Arctic tern, Devon Island.*

19 *Long-tailed jaeger, Bylot Island.*

20 *Young black-bellied plover, Bylot Island.*

1

21 *Arctic fox, Victoria Island.*

22 *Long-tailed jaegers are both predators and scavengers.*

23 *Nesting female ptarmigan on Bathurst Island.*

The arctic is vulnerable because of its individual nature, which is why the usual assessment factors that are applied to a tropical jungle or a temperate forest or a coral reef or a prairie grassland do not apply. All are vulnerable for different reasons. A tropical jungle is vulnerable, in part, because it has no topsoil. A temperate forest is vulnerable because it needs occasional pruning by spruce budworm and by fire, which we deny it. A coral reef cannot tolerate silting, or chemical pollution. A grassland is vulnerable to being overtaken by shrubs if shrub-eating animals aren't there. And so on. By its very nature, every environment carries the seeds of its own vulnerability. That is because every type of environment is unique in the way it functions. Ecology is specific to site and situation. It is not transferable.

Each environment specializes in the way it does things. Bacterial and other decomposing and recycling action is so rapid in the tropics that at any given moment virtually all biological material is locked up in the standing mass of plants and animals. The rains come and go on an almost rigid schedule, and things stay very much the same over long periods of time.

The arctic, on the other hand, specializes in fluctuation. *Fluctuation is the norm.* Oscillation is the way of life for every member of the community. Boom-or-bust is what lemmings and owls and snow geese and muskoxen are all about. Their vulnerability—and it is very real—lies not in their inability to deal with unpredictability, because that is their way of life. It lies in their probable inability to tolerate *additional* stresses with which they are not built to cope. They are stretched to the absolute limit already.

The polar bear population may or may not be able to tolerate the present level of commercial exploitation, but it is obvious to anyone that it could not tolerate pressure at the level of the old-fashioned African "sporting" safari. Bowhead whales have already proven, with all the other great whales, that they could not tolerate commercialization. Colonial seabirds are in the process of showing that although they can live with occasional losses of young, they probably cannot withstand losses of hundreds of thousands of adults to oil pollution and commercial fish-nets.

The well-known fate of the passenger pigeon in North America is a case in point. It has been estimated that at the time of the

European arrival on this continent there were between *three and five billion* passenger pigeons. By 1900 the bird was virtually gone and the last individual died in captivity in 1914. The obvious reason was colossal slaughter, but that does not tell the whole story. At some point, numbers seem to have been reduced below the level from which they could have recovered, even with protection. There was a population "threshold" which once crossed led to inevitable extinction for the species. The collapse was not sudden, however, but gradual and inexorable. Things in nature never seem to happen with sufficiently dramatic suddenness to satisfy our thirst for simple answers. In his monograph on the passenger pigeon, A.W. Shorger comments, "Every species of animal is doomed to extinction when it fails to produce sufficient young to equal the inherent annual losses. This the pigeon was not permitted to do."

Inherent losses in arctic wildlife populations are not necessarily annual. They might better be described as "occasional". But when they do occur, they can be devastating. The assessors of environmental impact should not make the mistake of assuming that the fine-tuning that has carried arctic animals and plants along through occasional natural "busts" can be expected to allow them to cope through stresses with which they did not evolve.

The Inuit of original days were also finely tuned to a very chancy existence. But they managed, not only physically but also culturally and psychologically. The ancient Inuit ways of doing things and of perceiving things was itself highly specialized, because it had evolved *with*—not in spite of—the arctic environment. Like the ways of all northern beings, those ways were survival-oriented. They worked. But they only worked so long as they were not tampered with by forces from outside that environment with which they had evolved.

Europeans arrived in the north in quest of profit. Profits were perceived variously in transportation routes, whales, or furs. The Inuit "natural economy", like the way of being of the tundra and the sea and the ice, was not geared to profit. It was geared to cycles and rhythms and oscillations which although not always predictable, were anticipated. The style of living, from lousewort to shark, from sculpin to bear, from hare to wolf to Inuit, was *appropriate*. Not consciously appropriate, of course, but adaptively appropriate. It worked. It had evolved on site, and it was perfect for that site.

The paleontologists and archaeologists in giving approximate dates to materials and events use the expression "BP" (before present) when dealing with a time scale that transcends the brief period of human history. In the arctic, however, BP means "before plundering". The arctic environment BP could clearly support a multitude of living beings of a good many species, including even a modest population of the human species. But because of its inherent unpredictability and its tendency to fluctuate, the arctic BP was very probably at its environmental optimum—the absolute limit of its tolerance—already.

Before we began to plunder it, the arctic land and sea and its plant and animal species had learned, the hard way, how to cope with a unique set of environmental conditions. And cope they did, through the miraculous flexibility, or adaptiveness, that is both the fuel and the direction of evolutionary process. Each year, thanks to the sun, the permafrost relented to the extent of a thinly thawed layer of soil for the plants that sustained the animals from mites to grizzlies. Each year, the icy covering of the seas gave way sufficiently for whales to breathe, birds to fish, walruses to dig for clams. Each year the ponderous yet glorious cycle turned in all its grandeur, all its majesty. Some years were not as good as others; some were downright disasters. But there was always a rich recovery year in store . . . so long as the system was left alone.

The rider of the high wire maintained his teetering equilibrium—swaying, adjusting, compensating. He kept on coping—gingerly, cautiously, but consistently. He did not fall. That was BP.

PART TWO
THE TIP OF
THE ICEBERG

Coast Guard ice breaker, Lancaster Sound.

Very generally the issues concerning the arctic are of two broad sorts: the "political" issues and the "environmental" issues. In full knowledge that there is a plethora of interpretations of both words, a brief explanation is necessary. Here, by "political" is meant not only the usual affairs of partisan politics, or of government and bureaucracy, but also the strategies, tactics and activities of interest groups on every facet of arctic questions. Very often the political issues that arise, whether in the making of policy or decisions, or in the conduct of formal or informal inquiries, or in legislation, or in environmental assessment, or in the evaluation of native claims, turn out to be questions of procedure rather than of substance. These procedural "road blocks" tend to generate more intrigue than they do open debate.

In our discussions, "environmental" is used in its literal sense of having to do with environments. Unless stated otherwise, it is confined to physical and biological environments. Many attempts have been made to capitalize on the "environmental movement" by describing as "environmental" any consulting activity that not so long ago would simply have been called a consulting activity. Here, issues in social, economic and cultural contexts are seen as being political, in the sense that they have to do with the human organization, not with the interests of nature. Thus, in this book "environmental impact" means impact upon the biophysical environments of the north. "Social impacts" are considered political issues.

But both political and environmental issues often spring from the same set of circumstances, and there is much overlap. The wave of oil spills on the high seas since 1967 has had obvious environmental consequences, but it also sheds light on the priorities of the national governments concerned. The practice of "environmental impact assessment" has important ecological limitations in the arctic, but also it reveals both the priorities and the strategies (sometimes even the tactics) of governments and corporations. Any major change in the distribution or the abundance of arctic animals will have profound influence on native northern people, both ecologically and politically.

It would probably be difficult to find more than a handful of "normal" Canadians who would come out four-square in opposition to oil and gas. Few of us hate oil and gas; indeed most of us, at least most of the time, depend upon one or the other as fundamentally and unquestioningly as we depend upon the breathable oxygen in the atmosphere. And we will for at least some time to come. What people seem to be concerned about are the *implications* of fossil fuel exploration, extraction and delivery, especially in dangerous and vulnerable areas such as the ocean floors or the north. Growing public awareness of the hazards, so long screened from us, of such activity has given rise to a great deal of public criticism of current "frontier" policies and practices, as demonstrated both by governments and by corporations, and by the apparent symbiotic relationship between the two.

Criticism, in popular parlance, is seen as the taking of a negative position, of damning something out of hand. In its real sense, however, criticism is no more than careful evaluation of something—a weighing of the pros and cons. A responsible critical review of a book, a play, or a movie may be just as positive in its conclusion as it can be negative. Usually the criticism falls somewhere in between. Critics of Canadian northern "development" programs and policies, apart from those (relatively few) who experience a doctrinaire "knee-jerk" in reaction to any private commercial activity, have attempted to weigh as best they could whatever evidence they have been able to unearth, toward a reasonable evaluation of the various dimensions of the question. (The reader should understand that the secrecy of the federal government with regard to the granting of leases and permits, the publication of research work and the development of policies has made informed criticism extraordinarily difficult.)

The most obvious danger to northern environments from the activity of the petroleum industry is contamination of the land or sea by oil spills. Although there is as yet little or no concrete evidence of a historical nature in arctic regions (we haven't yet experienced it), there is the undeniable record of oil spill disasters in southern regions. The best we can do at the moment is to extrapolate from these to the north, in some knowledge of peculiar arctic conditions.

OIL AT SEA

Pollution of the sea by fuel oil has been a chronic problem for wildlife since the first oil-fired ship was launched. And as long as there are oil-fired ships at sea, there will be misadventures. Most such misadventures cannot help but involve the release of at least some oil onto the surface of the ocean. That is assumed. No one in the shipping or any even remotely related business wants accidents to happen because accidents cost money. But they will happen, regardless.

On the other hand, there is a long and shameful history of the *deliberate* fouling of the oceans of the world by oil discharged from ships under way. This practice reached its zenith in the two decades following World War II. For many years, ships were in the custom of cleaning out their oil tanks at sea. Rather than do so in port, where idle time costs money, ships which had delivered a cargo of oil would put to sea again immediately, and would flush out their tanks with sea water while on their way to taking on the next load. Also, ships of all sorts are known to dump at sea both the oily waste-water from their bilges and the used oil from crankcases. Although both national and international regulations and controls have been enacted, the ocean is a big place, and "fly-by-night" shipping abounds.

Ironically, much of what we presently know of the impact of oil on seabirds, for example, was learned in the 1950s and 1960s as the result of these dumping practices. Much work was done both in Britain and in Canada; in this country it was chiefly through the efforts of the late Leslie Tuck of the Canadian Wildlife Service in Newfoundland that oil pollution of the sea came to the general attention. Tuck, whose monograph on the murres received world recognition in 1960, devoted a considerable part of his life to studying and reporting the impact of oil on the seabirds of the arctic, Labrador and Newfoundland.

In winter, vast flocks of murres from the arctic islands drift south on prevailing currents to waters off Newfoundland, where they find themselves on or adjacent to major shipping lanes between

North America and Europe. The concentration of the murres often coincides with the concentration of oil. If a bird's wings are oiled, it cannot fly, and if food is not immediately available, it will starve. Or, if its body is even slightly oiled, its feathers will lose their insulating properties and the bird will succumb to exposure in icy waters. Some birds, on the other hand, do make it to shore, where they attempt to preen their feathers clean. These will often die of starvation before they can take to the air or sea again, or will perish from the toxicity of the oil swallowed during the preening process.

It is impossible to know how many murres, eider ducks and other seabirds may have been destroyed in this way over the years. Tuck reported that the great majority of birds affected probably never reach shore. They die at sea, sink, and not even an approximate count is possible.

Such deliberate poisoning of the ocean surface will never be fully eradicated. There will always be a class of shipping in which expediency will dictate dumping, but it is probably reasonable to expect that by far the most serious threat to the seas by oil today is sheer accident. The record of accidental spills is cause not for mere concern but for raw fear. Oil tankers have become very large and numerous. At more or less regular intervals one of them cracks up.

The great modern series of oil disasters began with the wreck of the *Torrey Canyon* in March 1967. The tanker was travelling at about seventeen knots when she ran aground on the Seven Stones reef in the Atlantic between the Scilly Isles and Land's End, Cornwall. The first gigantic spill of our time was under way.

Spraying of the spreading oil with detergent, in attempts to emulsify and disperse it, was unsuccessful. (Some later believed that the chemicals used for this purpose may have been more toxic to the marine environment than the oil itself.) Subsequently the ship was bombed by the military in an attempt to release the remaining oil trapped in the hull, which was then set alight and burned off with aviation fuel, napalm, and such material. It took about two weeks for these and other methods to dispose of most of the oil in the wreck and in the immediate vicinity. By that time, 160 kilometres of the British coastline had been oiled; it was estimated later that 25,000 seabirds died. It was a good ten years before the local biological system appeared to have healed.

73

Few Canadians can have yet forgotten the *Arrow* disaster at Chedabucto Bay, Nova Scotia in February 1970. When the ship came to grief she was carrying almost nine million litres of viscous Bunker C oil; the spill fouled 300 kilometres of coast, only fifty of which were cleaned up. Except for a small patch of shoreline which was selected for experimental purposes, dispersants were not used here, not being effective for this type of oil. In the absence of any contingency plan, attempts were made to pump oil out of the wreck, but the clean-up effort finally boiled down to backbreaking work with shovels, garden forks, rakes, and bare hands in efforts to clear the sticky mess. Bulldozers and front-end loaders were also brought into play, and an endless-belt skimmer called the "slick-licker" had to be invented on the spot. There was no emergency "drill" of any kind; local people simply did the best they could without any prior planning. The loss of birds, fishes, shellfish and other marine (especially intertidal) organisms could not be reliably estimated.

Then there was the *Amoco Cadiz* and what she wrought upon the coast of Brittany in the spring of 1978. In this case the gigantic tanker (like the *Arrow*, of Liberian registry) spilled over 200 million litres of crude and bunker oil over a horrific fifteen-day period. She oiled 304 kilometres of the French shoreline, killed birds and fishes and marine invertebrates, and contaminated entire coastal salt marshes. High winds carried the oil farther inland. No one yet knows the full extent of the damage.

Again clean-up efforts were primitive. People resorted mostly to mechanical collection with shovels, rakes, and pails. Some conventional recovery systems were used, but the most effective means turned out to be sludge pumps and septic vacuum trucks. In deep water, some detergents were applied. Amoco later charged that the French spill control personnel had misused the dispersants. Whether or not that charge was ever borne out, it was clear that once again technical anticipation and preparation for clean-up were not adequate to the enormity of the task.

The reader may also remember the good ship *Kurdistan*, of British registry, which in the early spring of 1979 oiled over 1200 kilometres of Nova Scotia coastline and another 105 in Newfoundland. Over five and one-half million litres of dispersant-resistant Bunker C oil were spilled—*this time in the ice*—and it did not surface until warmer weather came. The attempted clean-up was, once again, primitive and labour-intensive, because the thin wide-spread slicks that eventually surfaced resisted conventional mechanical means of capture. Once again, shovels, rakes, and garden forks were used to load oil and oily debris into drums and plastic garbage bags. This was the first "real life" (as opposed to experimental) experience of the formidable combination and interaction of oil, water, and ice.

There seems no option but to expect that there will be more such events as supertanker traffic intensifies on the high seas of the world. Year-round shipping, under all conditions, is being seriously proposed for the Northwest Passage. Accidents are wholly unpredictable as to timing and location, but entirely predictable in the sense of probability. The major difficulty with ship breakups is that you cannot anticipate where in the world they are going to happen. Collisions can be anywhere, and reefs and shoals are worldwide. On the other hand there are accidents such as drilling rig blow-outs which are equally predictable in the sense of probability, but at least we know that these are not going to happen where there is no drilling going on. To the extent that we know where drill rigs are, the geographic focus of concern can be, superficially, narrowed.

To those who explore for gas and oil, there is no dread so deep-seated as that of a blow-out—the sudden release of immense subterranean pressure at an unexpected moment. In the high arctic islands, unfortunately, and in the Beaufort Sea and the Mackenzie Delta there are zones of "abnormally high" geostatic pressure, which of course heightens the possibility of accident. In the summer of 1969 the first such event in the Canadian arctic took place at Panarctic's Drake Point drilling site on Melville Island. It took two weeks to shut off the gushing gas. A month later the well blew out of control again; this time it could not be shut down until more than a year after the initial explosion. During that time it lost gas at the rate of about 85,000 cubic metres per day. Five months after the Drake Point well first blew out, another Panarctic well, this time on King Christian Island, went up. This one lost gas at about 2.8 million cubic metres per day, and, unlike the Drake Point gas jet, this one was on fire. The gigantic flame was finally extinguished three months later.

Observers asked at the time, and have wondered since: what if these blow-outs had involved not gas, but *oil?* Oil does not burn "cleanly", and it does not disperse in the atmosphere; it collects on the surface of the water and under the ice. It does not go away. To date in 1980 we have not heard of an oil blow-out in the arctic, but we do have the example of the colossal 1979 disaster in the warm subtropical waters of Campeche Bay in the Gulf of Mexico.

On June 3, 1979, ixtoc 1, an offshore well of Pemex, the Mexican national petroleum company, blew out of control. For months, it spewed oil at a rate of over one million litres *a day.* By the beginning of November more than 375 million litres had poured into the Gulf. Some of this was burned off—what proportion is not known—but the greater part is thought to have entered the water. Virtually every control and recovery device known was tried at one time or another. The flow was eventually stopped, but not before incalculable damage had been inflicted upon the marine biosystem by the largest oil spill in history. It may be years before the extent of the impact can be estimated. Perhaps the most important lesson of the Campeche catastrophe is that it was so extraordinarily difficult to contain, even though it did happen in one of the more benign regions of the world, so far as weather is concerned. No one likes to imagine what might have happened, or how long it would have taken to plug the hole, or how widespread and how lasting the effects might have been, had such a blow-out happened under arctic conditions.

Leaving to one side for the moment the sheer mechanical difficulty of dealing with an oil blow-out under the ice, or on the sea floor, or on the permafrost, the possible consequences for wildlife such as seabirds, even on open arctic water, are hair-raising. (Of course a blow-out of Campeche massiveness would not be required; a much lesser spill, or even the accumulated effect of "normal" leakage, could create havoc in high arctic waters.) David Nettleship of the Canadian Wildlife Service estimates that a major spill in the latter half of August could literally remove "as much as three-quarters of the total murre population in Lancaster Sound and Jones Sound." Just then, the birds are gathering on the water for their fall exodus. Supposing that the murres had experienced a poor reproductive season—which happens—one can only ask, what then? Coincidentally, the arctic weather cycle is such that the drilling season is short, and usually drilling takes place in August, just the time when at Coburg Island, for example, the murres are flocking at their highest density for migration.

THE RISK FACTOR: ONE IN A MILLION?

Carson Templeton of the Winnipeg-based Environment Protection Board put it to *The Nature of Things* succinctly: "As long as people are doing things, things are going to go wrong. And Murphy's Law—if something can go wrong, something will go wrong—certainly applies in the arctic, where there are very difficult circumstances." The critical question seems to be this: in certain knowledge of the undeniable risk, is that risk worth taking?

Many say that it is. Given the notorious uncertainty and fluctuation of the arctic system, including the weather, and its stubborn resistance to prediction, some believe that rather than trying to anticipate what might happen, we should forge ahead in the best way possible and learn as we go. That is reported to be the approach Norway has taken in the North Sea: rather than attempt to predict the precise circumstances, they have chosen to go ahead without preparatory studies, but they will have set up the best possible permanent monitoring system. In this way, when something does happen, at least they will be in a position to examine and follow it from the very beginning. This would be the "hands-on" way of learning, but it would be practical, not theoretical.

Naturally such an approach must assume that there will not be an oil spill of sufficient size to do lasting environmental damage, or, that if there is a serious spill they will be able to contain it. As we have seen, experience in other parts of the world indicates that neither of those assumptions inspires confidence. On the other hand, many of the most responsible scientists might say, quite correctly, that this would be the only way to learn *for certain* about

77

the behaviour of oil in northern seas. Small, contained experiments simply are not sufficiently informative—which is probably accurate. It all boils down to the acceptability of the risk.

Industry clearly has no choice but to be positive and optimistic. According to William Henry, president of Norlands Petroleums, "The key is that it is not going to be easy, and it's not going to be done cheaply. It's going to be an expensive proposition; it's going to require more equipment, more personnel, and the application of better and up-to-date modern technology. We have drilled all over the world and in many hostile environments—in extreme cold, as in Siberia, where there are substantial oil and gas reserves—and we have built pipelines there. The evidence is that it can be done, and done safely. We've done it in jungles and hot desert areas, where again you have sensitive environments. It's a matter of applying the right technology to fit the situation. I am convinced, to date, that the evidence is there to show that the technology is capable of handling those situations."

In any event, quite apart from the pervasive contemporary faith in high technology, industry spokesmen have long insisted that the risk of a blow-out or a major spill of oil is "one in a million", perhaps even less. Henry pointed out that "normally, in most exploratory operations like this, we see very little [in the way of] blow-outs because of extra precautions that are taken when drilling in new areas." This does not jibe with Panarctic's gas blow-out experiences on King Christian and Melville Islands, where exploratory drilling was done on a "wildcat" (probe and hope) basis. Judging by the length of time those wells ran out of control, precautionary measures, if they existed, must have been less than reliable.

The Canadian Arctic Resources Committee has thought long and deeply about the "one in a million" claim. CARC's Donald Gamble says the claim is "nonsense", and that the chance of an accident in Lancaster Sound, for example, would be "a statistical certainty" were exploratory drilling to lead to production. He bases his judgement not only on our experience in the south, where environmental conditions are so radically different from those in the arctic, not only on the unknowns of exploration in new areas, not only on the high-pressure spots in the north, but also on the *later* events to which even safe and successful exploration would eventually lead. "If we go into production, we have sea-bed pipelines, we have pipelines coming ashore to tank farms, we have ports. . . People in government, in the Arctic Marine Oil Spill Program, have said that if we develop a full-producing hydrocarbon province, if we develop this area as a major transportation corridor for oil tankers, we will have accidents."

Interestingly enough, the record seems to show that in spite of the existing blow-out record, the exploratory drilling phase is the *safest* part of fossil fuel production. Although the Campeche blow-out was the greatest spill in history, most of the major disasters have involved tankers. In any event, the "one in a million" estimate is challenged on many grounds. In Alaska it is reported that one in every two hundred wells results in a blow-out of some kind—not a major one, admittedly, but still a blow-out, and the scale can increase very quickly.

Carson Templeton, himself an engineer, told *The Nature of Things* that in any case the "one in a million" figure concerns the *first* well: "You don't expect the first well to get oil anyway, so therefore the risk is slight." He added that in view of recent risk estimates, ranging from one in a hundred downward to one in a million, his own judgement of the odds would be "somewhere from one in two hundred to one in a thousand." As a matter of record, three of Dome Petroleum's first ten exploratory wells in the Beaufort Sea experienced water blow-outs, which involved the loss of pressure, not oil or gas. We do not hear of the numerous minor events that take place in the north because they do no apparent damage. But to those who are skeptical of the "one in a million" risk estimate, these little happenings add up to a chilling portent.

The question that must now be faced, everyone acknowledges, is this: even granting that the danger of a major oil spill might be statistically "low", if one *were* to happen, what then? Could we contain and control it, or couldn't we? Failing containment or control, could we even clean up afterward?

CLEAN-UP: THE STATE OF THE ART

Few, even of the most positive-minded and sanguine members of industry and government, would claim that there is yet the technology to clean up a significant arctic oil spill. When interviewed, Gerry Glazier of Petro-Canada freely acknowledged that "we all feel a little nervous about the technology of cleaning up oil in the arctic. I don't think anyone can pretend to say that you could successfully go and clean up a major oil spill off the surface of arctic waters, because it is obviously a hard environment to work in. The weather is certainly going to be a big factor, and there is a lot of ice around. I think that we have to start looking at other ways—just physical clean-ups, that could lessen the impact of the oil." This seems to say, with complete forthrightness, that we will just have to do the best we can.

In a 1979 interview Jake Epp, at the time the federal Minister of Indian Affairs and Northern Development in the Conservative government, put it even more bluntly. "If you are asking me, 'Do we have the technology?', I think the simple answer at this stage is 'no'." He did however add a curious footnote. "We have learned a lot through our experiences in the Beaufort, and that experience—that knowledge—is transferable." Presumably Epp was speaking of drilling technique, not clean-up technique. Our experience of a major oil spill in the Beaufort Sea is zero, I am happy to relate. There has been nothing to clean up, nothing to learn, and nothing to transfer. Dome Petroleum, to its credit, has mounted a massive research program on oil spills. But the millions invested have yet to yield any real answers. There is a lot of optimism and a lot of hope, but the fact remains that the state of the art is grossly out of step with the risks that are being taken.

The oil companies appear to be banking on preventative rather than curative measures. By drilling with the utmost caution and making use of all the most recent developments in technical know-how, they will hope to avoid the clean-up problem by forestalling accidents before they can happen. William Henry of Norlands said, "Part of the studies that we have been doing are looking at ways to minimize any effects from [accident], and certainly to drill in a manner where we don't have that kind of situation develop in the first place. . . The studies that we are doing will tell us during what time of the year it is safe to drill, so that we can move equipment in and out. . . . Because of the great logistics you run into, in moving and having equipment available to you, [it] runs the cost up considerably, but that's part of maintaining the concern so that we don't run into situations like they have run into in the Gulf of Mexico."

Indeed Henry is one of the very few who will go so far as to say that if a spill or a blow-out *should* happen in the north, he feels "very comfortable" about the ability to handle it. As he explained: "We have made studies and are continuing to make studies in how to handle it. We have talked to many experts, and there is a group of companies. . . that have put together a group to assemble material, equipment and personnel to handle a situation like that, should it arise. . . With the technology of booms and equipment able to pick up the oil, or contain it, you have an even better chance today to control those kinds of situations if they occur. I think that our evidence so far indicates that the same kinds of procedures that were used in other areas of the world can be done safely in these arctic waters."

Not so, says Donald Gamble of CARC, if we have learned anything at all from the spills that have been experienced in the south. "We know that we cannot handle those in a relatively temperate climate. You can just imagine the difficulty we will have in the ice-infested waters of the north, even at the best time of the year, in midsummer. If a blow-out occurs late in the season, then we have ice cover and pack ice to deal with."

When the government of Canada approved offshore drilling in the Beaufort Sea in 1974, it recognized that we did not have the capability for cleaning up oil. So the Arctic Marine Oil Spill Program was set up; its specific task was to try to find a means of cleaning up. After a great deal of study, it was concluded that it would be possible to burn off spilled oil. "Experience in the Gulf of Mexico," said Gamble, "has shown that that is not possible, so we are right back where we started from. Interestingly enough, the Arctic Marine Oil Spill Program budget was cut by 50 per cent [in 1978], so here we are. . . pulling out the technological back up." Even as we continue to argue about whether or not *some* of the oil can be burned off.

As in so many other of his daily activities, *Homo sapiens* seems to be able to draw startlingly different conclusions from the same set of evidence. Either we learned something from the *Torrey Canyon*, the *Arrow*, the *Amoco Cadiz*, the *Kurdistan*, and especially from Campeche Bay, or we didn't. It appears that different people learned different things.

Whatever was or was not learned will very likely soon be put to the test in the Canadian arctic. In the meantime, while discussion of the technology of high latitude clean-up goes on, so too continues the exploration of the north, apparently quite in spite of the weakness of the present information base. It continues because of the insistence of (especially) the federal government but also of industry that we *need to know* what is down under that ice and water and permafrost in the way of oil and gas. The "need" is expressed as an absolute and unchallengeable imperative, with a solemnity and gravity that is usually reserved for pronouncements from the Mount.

THE NEED TO KNOW

The rationale for the government's introduction of the policy in 1976 was quite simple: as a nation, Canada needed to know what inventory of oil and natural gas it had. Otherwise there could be no informed discussion on such matters as energy policy. If we did not know the extent of our fossil fuel reserves we could not make intelligent or reliable decisions. This was eminently fair and reasonable. But it went further. The next step in the "need to know" program was to make a huge investment of public funds in the building of that knowledge. This was done by way of tax incentives for those companies that undertook exploratory drilling.

Many countries do give companies permission to search for fossil fuels, and they pay the companies when they make finds. They do not however give the companies either ownership over the oil and gas in the ground or the mandate to market it. In effect, such countries reserve any marketing decisions to themselves. They review the inventory thus established and then decide if and when the fuels should be extracted, and by what means they should be marketed. Clearly this approach to appraisal would allow careful and deliberate thought to be given to the extraction of oil or gas in situations which might be environmentally hazardous.

In Canada, however, it was different. Andrew Thompson, an oil and gas expert and chairman of CARC, described the difference. "The implication that is held out when they talk about the 'need to know' is basically not honest. It is well demonstrated that the companies that commit their money to an exploration venture do it on the undertaking that if they make a commercial discovery they will be able to exploit it."

The tax incentives that were established for oil and gas exploration in the far north were so attractive that a drilling company that was suitably financed could scarcely afford to refuse them. Indeed, as Gamble put it, the companies were effectively "compelled" into frontier regions like the arctic. We will return to finances in a moment. Here it must be emphasized that the "need to know" policy was tantamount to a first step toward the actual extraction and marketing of fossil fuels by private corporations. On the evidence, there is no other way in which to interpret it.

Although the federal government will insist that a permit to explore is not a permit to extract or to transport fuel, Carson Templeton sees it as the old "foot-in-the door" technique. "Any salesman will tell you that you get your foot in the door first. Then you [the government] say, 'Look—here is something that is going to keep you from freezing in the dark,' in the hope that that will generate support for it. And then you go on to the next and the next. And once they have spent a hundred million dollars, you say, 'Well, we can't throw away a hundred million dollars. We've got to go ahead.'"

The essential element in the federal government's "need to know" policy was that companies such as Dome Petroleum, drilling offshore in the Beaufort Sea, and the various others elsewhere in the arctic, could in fact *make* money from oil and gas exploration, by the simple expedient of writing off as much as 200 per cent of their costs. And if oil was found, the company kept *all* of the profits.

MONEY

In the fall of 1980 the federal government brought forward a "national energy policy", the final form, the full implications, and the actual effect of which all remain to be seen. The "super depletion allowance" had simply become too embarrassing, so it was jettisoned. What will effectively take its place was not known at the time of writing, except that a new system of exploration grants will be tied to the degree of Canadian ownership of the companies concerned. The larger grants will go to those that are "most" Canadian. There was an immediate reaction from the oil industry against this apparent reduction from their tax savings, with the loudest protests coming from the foreign-owned companies.

Quite naturally, the oil exploration company sees its own money as its own money, and the fact that it may receive a 200 per cent tax break does not change the fact that it has indeed invested its own money. That the Canadian taxpayer must foot the bill is part of the corporate calculus. Corporations are in business to make money.

The Canadian taxpayer foots the bill for arctic exploration by the simple expedient of being required to make up the difference between revenues forgone by government and the total tax budget. You don't pay the companies directly; you pay government its cost of funding exploration. Also, as energy commentators have pointed out, you pay additionally as a consumer. In this case, as well as meeting federal and provincial taxes, you do pay the companies directly. It has been calculated that you are presently paying prices that are at least ten times the original cost for the (relatively cheap) oil and gas that was discovered from the 1940s to the early 1960s. Finally, when the generously subsidized oil exploration company does make a discovery today, it of course feels justified in charging you, the consumer, the world (or oil cartel) price for the new fuel. And as we have pointed out already, the company feels that it has a prior right to the extra profits.

Of course the oil companies (the largest of them foreign-controlled) are unanimous in their claim that without the government subsidy they receive in the form of tax incentives they simply could not afford to maintain the level of exploration that is required. Hence their sense of outrage at the 1980 energy policy announcement. Even in the face of unprecedented recent profits, those profits must be re-invested in order to maintain the pace of exploration. This brings us back to the "need to know"; the argument is all of a package. It should be added that in spite of the exploration re-investment imperative, hundreds of millions of dollars are paid to parent corporations in the United States by their Canadian subsidiaries. Richard Gwyn of the Toronto *Star* reported in August 1979, "Shell Oil last year so ordered its affairs that it paid not one cent of federal taxes." CARC is quick to point out that Dome is in a similar situation.

Walter Gordon, former Liberal Minister of Finance, observed at a 1975 meeting of the Canadian Arctic Resources Committee that "some Canadians do not trust the leaders of the oil companies. . . . Canadians are skeptical about statements by the government, by government officials, by the National Energy Board, and last, but by no means least, by spokesmen for the foreign-controlled oil companies in Canada. . . . I believe control of the foreign-controlled oil and gas companies in Canada should be transferred to Canadians. . . . Failing that, I think these companies—or certainly Imperial Oil, the largest of them—should be nationalized." The federal energy policy of 1980 suggested that someone may have at last been listening.

Not that it matters much to a Lancaster Sound narwhal or to a North Slope caribou or to a Beaufort white whale who pays the tab or where the profits go. What will matter to them will be an oil spill, or a blow-out, or the more gradual and less discernible but still inexorable "destruction by insignificant increments" of the natural system that sustains them. But it may matter to at least some Canadians when they eventually become more fully aware of the enormous complexity of government-corporate fiscal inter-relationships and how these in turn affect the north and its inhabitants. Quite simply, the arctic is being penetrated at breakneck speed by companies that are making enormous profits; you are paying for it; and you are paying for it most handsomely indeed.

It is possible, as the public knowledge and understanding of the sources, the uses, and the disposition of oil and gas funding and profits becomes more sophisticated, alternative applications of at least some of those monies might be openly discussed. A mere fraction of the public tax dollars now flowing through the Beaufort Sea, for example, would be a bonanza to those studying solar energy technology, or to those seeking to develop and refine techniques for efficient energy conservation. Research in alternative energy fields is fairly starving for government support. On the basis of some recent revelations of oil corporation investments, by the way, there is some reason to think that some of those profits are going not into more exploration or even to foreign parent corporations, but toward investment in those very energy options that will one day soon replace non-renewable petroleum fuels. As it tends to be in all places, irony is rampant in the north. It would be depressing to think, perhaps, that the devastation of any part of the living arctic would have to be the price of tar sands development, or of your solar greenhouse, or of the synthesis of burnable fuel from pig manure. But it just might.

ENVIRONMENTAL ASSESSMENT AND REVIEW

We cannot know at this time what if any part of the Canadian arctic will or will not be devastated. We can only guess. We can improve our guess with knowledge (very often surprisingly difficult to obtain) of the localities and the nature of exploratory and other industrial activities in the north. To this basic information we would attempt to apply such knowledge as we have of, for example, arctic biology, ecology, geology, oceanography, meteorology, and so on, in order to arrive at a prediction of what is likely to happen, both over the short and long terms, at a particular site. This prediction we would call an "environmental impact assessment" or EIA. (This would be a biological or ecological EIA; there are of course social, including economic, cultural, and other assessments.) In their *Environmental Planning Resourcebook*, Reg Lang and Audrey Armour define EIA as "a form of pre-active [before the fact] evaluation intended to determine whether to proceed with a given project and/or how to proceed so as to prevent or minimize environmental degradation."

Everything hinges on the reliability of the data that are fed into that evaluation, and on our competence to project conclusions from those data into the future. The depth and thus the reliability of our knowledge of arctic ecology has been touched upon in Part One. In general although a good deal of excellent *baseline* (descriptive) biological and other work has been done in the far north, it would not be an exaggeration to say that our understanding of arctic *ecology* in the "process" or systemic sense is very low indeed. As pointed out in Part One with respect to the marine biology of Lancaster Sound, *we have not been asking the most basic questions.* We are very good at asking "what?", not so good at asking "how?" and we almost never ask "why?" We spend a disproportionate part of our time in simple description:we rarely attempt to tackle relationships. Perhaps it is just as well that we do not, because we most certainly do not know how to begin to marshal the answers. If indeed ultimate "answers" are ever to be expected.

Max Dunbar, who was quoted earlier on the open questions concerning the biological productivity of Lancaster Sound, in a 1979 keynote address to the CARC-sponsored symposium on Marine Transportation and High Arctic Development, had this to say about EIA:

> the most important requisite is basic research, something that should have been obvious from the start. There is a school of thought that believes that ideal impact studies, successful in predicting accurately the result of accidents and industrial wastes, may well be impossible. Nevertheless, it has at least become clear, even to the most refractory minds, that, in order to come even close to the ability to predict such effects, it is necessary to know precisely and simply how nature works in the particular context at issue. What is needed is basic science, not "integrated, interdisciplinary, mission-oriented" jargon.

The "credibility" of EIA has been sorely frayed since it found itself surrounded by the kind of jargon Dunbar cites. And for good reason. Rarely in our time has an enterprise which originated with the best of intent and for the best of purposes been so beset and afflicted both by unabashed opportunism and by brutal cynicism. (For more "free-wheeling" discussions of EIA, the reader is referred to the Selected References and David Schindler's article and this author's 1981 *The Fallacy of Wildlife Conservation*.)

EIA is obviously as good as its data, and as good as our ability to interpret those data in some predictive way. In the arctic at least, these are stern limitations at the present time. As we have seen, the essence of northern terrestrial life systems seems to be unpredictable fluctuation. It may well be, as our knowledge strengthens, that the marine system will be more susceptible to our understanding than the land, if only because of the relatively more "stable" nature of marine environments. We cannot yet know that. Ecological prediction in the arctic is a very dicey affair. One would hope most fervently that the fate of any part of the marine or terrestrial arctic will not depend on how some committee of technocrats interprets the primitive knowledge available at present.

There is another difficulty with EIA, and this one is not particular to the arctic. This has to do with its *style*. Since it must appear to be "scientific", an environmental impact statement usually quite self-consciously assumes the appropriate stance of clinical "objectivity", complete with the language of quantification. This, especially where the information base is tentative, may give the impression that the material presented and conclusions drawn from it are more reliable than they actually are.

This phenomenon was observed in the United Kingdom by Ronald Bisset, who points out a number of ways in which the style and language of quantitative methods "may avoid or weaken opposition to a proposed development or policy." First there is projected an apparent visual and arithmetical complexity which gives "the impression of scientific accuracy." Laymen, including elected representatives, might well be persuaded that "environmental, social and economic sciences are able to predict accurately the impacts of development." This may well inhibit questioning. "Should a written account be presented, the quantified data might be treated as an exact representation of the information in the submission and particular attention may not be paid to written evidence. Thus, the use of quantitative methods may convince those

88

who are not experts that the quantified representations are accurate." In other words, skilled *presentation* may have a greater influence on lay reviewers (which means most of us) than actual content.

Even more serious is Bisset's next point. "Those who choose the method of assessment are able to make a strategic choice which may give them a tactical advantage when pursuing a particular outcome. Opposition groups or individuals may have to frame their views in terms of the method used and might challenge only particular quantitative representations." In other words, attention can be diverted from certain aspects of the EIA and directed toward others. And of course specific methods can be selected with an eye to anticipating opposition.

In 1973 and 1977 the federal Cabinet directed the establishment of the "Environmental Assessment and Review Process" (EARP), calling for guidelines, environmental impact statements, review panels of experts, public hearings (as appropriate), and final reports, having to do with proposed projects which may have environmental implications.

A recent EARP involves Norlands Petroleums Ltd., a wholly-owned subsidiary of a U.S. utility that, with Petro-Canada and Shell Oil, holds oil and gas concessions to approximately 3.2 million hectares in Lancaster Sound. A drilling permit has however not yet been issued. In June 1978 Norlands published an "Environmental Impact Statement for exploratory drilling in the Lancaster Sound region." Hearings were held, but a decision was deferred and remains in abeyance. In this case the EARP recommended a "comprehensive review of the complex resource use problems in the Lancaster Sound Area... by the Department of Indian Affairs and Northern Development" and the preparation of a Green Paper (from which this is quoted) to provide a "meaningful assessment of environmental and socio-economic impacts of exploratory drilling...."

A special twist to the EARP is that the proponent of a project, while (quite properly) being required to pay for the environmental assessment, also carries it out. Guidelines are issued by the government, and the company must meet them, but how the assessment is actually done is the company's responsibility. The eventual state-

89

ment is then scrutinized by government experts, and judgements are made. In most cases, decisions are in the process of being made on the basis of a very limited *project-specific* (remember, fragmentation is the style) EIA done by the oil and gas industry itself. (As David Suzuki remarked at one point on the TV program, this is rather like asking the tobacco company to assess the potential hazards of smoking.)

To offset any disquieting implications that may arise from all this, the reader is referred to Milne and Smiley's careful and revealing *Offshore Drilling in Lancaster Sound: Possible Environmental Hazards* published by the Department of Fisheries and Environment in 1978.

It is a complicated affair, EIA, and when we explore beneath the surface, not a comforting one. And, as Carson Templeton reminds us, ". . . the EARP process is not automatic. The Minister has to say what other things he will consider, and what will be put to the EARP process. So things like the Dempster Highway or the Mackenzie Highway, government-oriented things, are never subjected to an EARP hearing. The drilling in the Beaufort Sea . . . was never put to an EARP process. The public had no input whatsoever in that decision. In fact, the Minister approved it after midnight in 1976 just as Parliament was going to recess for Easter break, so that he wouldn't get any comments." The reader is reminded of the definition of EIA: it is an evaluation intended to determine *whether* to proceed . . .

Not merely, then, is EIA only as good as its data; it is not much good unless it is required for a project. Assuming that it is actually carried out, however, it is only as good as its terms of reference. Here we return to the fragmentation problem. Don Gamble says, "what you have to understand is that the assessment that a consultant is asked (by the company) to do is usually very narrow. He is asked to do a short-term study, the object of which is to gain approval for the project . . . In the case of Norlands, they were asked to examine the significance of a single exploratory well. But you never explore a geological structure with just one well. We all know that there will be a great many wells, before the company either gives up or succeeds. Theoretically, they are supposed to go through the assessment every time. But once it is set in motion, the treadmill is very hard to stop. They will use the previous decision to justify continuing. Later, they will say, 'Well, you allowed us before; we have spent millions and millions of dollars now; you have got to let us carry on,' in spite of whatever new risks are brought up. This is what has happened in the Beaufort Sea." Another instance of the "foot-in-the-door" metaphor.

A still further difficulty inherent in EARP is the quality of review to which an impact statement may be subjected. Not the quality of the review personnel, but of their review. Unfortunately, government sponsorship of primary research has been grievously reduced in recent times, with the result that the greater proportion of the "newest" information rests in the hands of industry-sponsored private consultants. There may not be government or other "experts" who are sufficiently up-to-date to review it. No one would deny that a consultant has a professional relationship with his client that may well involve confidentiality. However, it is not of much help to an EARP if the reviewers are not privy to all of the appropriate information. It would be a severe limitation to the questioning process.

Theoretically there would be a further EARP were a company to actually find gas or oil and wish to commence extraction. This has not yet come to pass in the arctic; no one has yet gone into full-scale production. However, there are signs. As in the Beaufort Sea, once the government has approved initial exploratory activity, that activity tends to expand far beyond what anyone may have expected at the time of approval. And of course there are drill sites in the central and high arctic that never underwent any formal review at all. As was pointed out earlier (in the discussion of "need to know"), once an exploratory program is allowed to go ahead, a decision has already been made that, if something is found, production will follow. It is important to remember that the original decision was made on the basis of *one* exploratory well. The momentum of the treadmill is very real.

This is what is meant by "destruction by insignificant increments". As we have seen, the last things a project proponent wishes to discuss in his EIA are cumulative, incremental, or synergistic impacts. The simplest and most expeditious way to deal with his material is to treat each aspect of the proposed operation and each

91

aspect of the impact study in isolation. Here, if he is so disposed, a proponent can fall back quite comfortably on the limits to ecologic predictability, and a consultant can quite properly decline to make what would be scientifically irresponsible projections of a quantitative sort. Also, if the terms of reference of an initial EARP are concerned with one exploratory well, who is to say if oil will *ever* be discovered, much less what an extracting and processing facility might look like, still less what form a future transportation system might take? All that would be outside the mandate of the EARP.

Another aspect of this is our general blindness to gradual, subtle change. In a very real sense this is part of our innate human "adaptiveness": we can get used to almost anything—even ageing—over time, and never really notice the changes that have been occurring. (For further discussion of this the reader is referred to the author's *One Cosmic Instant*.) Reg Lang and Audrey Armour stress the insidiousness of the process in urban environments: "The environmental quality of urbanizing areas is lost by attrition, and people tend to adapt to environmental change that is slow and disjointed. . . . It is a slow, bit-by-bit process of attrition, hardly noticed and seldom monitored." One day you wake up, look about, and see that *everything* has changed—not merely in degree, but in kind. You are in a totally new situation. This is not an individual or urban phenomenon only; natural environments can change also, from forest to desert, from healthy to unhealthy, from clean to foul. Like the extinction of a species, it happens gradually, and by the time you notice it, the process has usually passed the point of no return. The change has been *qualitative*.

Environmental impact assessment, as we presently know it, whether it is used unthinkingly, casually, carelessly, or downright cynically, can make a devastating contribution to this process. It *need* not do so, but it *can*.

SOCIAL IMPACT

Although reliable assessment of the ecological consequences of industrial activities in the arctic has scarcely begun, the social consequences for native people of the white presence in Canada have been visible since the seventeenth century. Such knowledge, you might think, would make the prediction of "social impact" a great deal easier than the prediction of ecological impact; after all, the results of what we have accomplished in the north are already highly conspicuous, and have been for a very long time. In spite of this, however, we hear of a mounting volume of new "studies" meant to reveal what may happen next. It all seems strangely myopic.

Leaving to one side the many negative criticisms that have been levelled at "social impact assessment", and also leaving to one side the question of whether such study is even needed in view of the overwhelming evidence we already have, let us consider for a moment the implications of the contemporary predictive approach. As we have seen, environmental impact prediction is required to be quantitative: it must express, in numbers, what the effects of a certain course of action may be expected to be. It turns out in the final analysis, however, that the effects of accumulated impacts over time are not quantitative but qualitative. The destruction or radical transformation of a biological community cannot be expressed in numbers because the change is in kind, not in degree. The same is true of social impact; the destruction or radical transformation of a human society or of a culture represents the crossing of a threshold that can never again be experienced. There is no going back. The crossing of that threshold is *final*. It is extinction, and from extinction there is no appeal.

The "multicultural mosaic" that is the trumpeted policy of the Canadian government cannot without profound re-thinking meet the requirements of indigenous cultures. The reason is quite simple. Canadian "multiculturalism" as advertised is *in fact* an aggregate (sometimes a "mix") of the cultures of the *Old World*—British, French, German, Portuguese, Ukrainian, Mediterranean, Asian, and all the rest—overlain with the more recent changes (adaptations, modifications, mutations) that have arisen uniquely in rural and urban North America. It does not include the cultures of the original peoples. It never has and it never can accommodate indigenous cultures because those cultures are not industrial.

Canadian "multiculturalism" is in fact international industrial monoculturalism.

In his Mackenzie Valley report, Thomas Berger remarked that when they observed the native people's use of the land the "Europeans had no difficulty in supposing that native people possessed no real culture at all," because "European institutions, values and use of land were seen as the basis of culture." This is of course a perceptual problem, one to which we shall return in the next Part. Here it should be pointed out, however, that over the generations those white Europeans have become Canadians, but even after this passage of time the industrial "developers" and their natural allies in government and bureaucracy still appear unable to perceive that there is any culture or any ethos beyond that of industrial production. We tend to pin a disproportionate measure of the accountability for this on our forefathers; it is obvious that the general failure to *perceive* northern cultures remains by far the most stubborn of our "multicultural" blind spots. Nothing exists beyond the monoculture; how can "nothing" have a claim, or a vested interest?

It is far too easy for the do-gooders (which includes all of us) to focus our attention on the nitty-gritty of native jobs, alcoholism, family problems, urban displacement, legal niceties, even claims and "rights", while never addressing the *real* issue, which is the systematic and indifferent dismantling and liquidation of a culture far older than the industrial monoculture, far wiser, and far more adaptive. Since in the urban-industrial monoculture money is the answer to everything, we translate every problem—except the real problem—into dollar terms. Budget enough money to native "rehabilitation" (which actually means "assimilation") and the problems will be solved. We overlook the radical truth that since cultural extinction is final and irretrievable, it is not measurable. You can't put a price on it.

The urban-industrial monoculture, the melting-pot that is the real as opposed to the expressed policy of the Canadian federal government reveals "multiculturalism" for the travesty it is. Clara Schenkel of the Yukon Indian Cultural Education Society has written, "In theory, the melting pot concept is supposed to produce a new and happier way of life for all its citizens. In practice, particularly in North America, it has produced an intense and deepening materialism." For the native, she says, integration "means adopting a set of values against which his deepest nature rebels." In northern communities we see the result of this conflict in all its familiar manifestations. We like to find the cause of native "apathy" in native "nature"; we are unable to recognize its roots in the wanton and deliberate eradication of an appropriate culture by the white urban-industrial ethos—so inappropriate, as we have seen, to northern conditions.

At the northeastern corner of Baffin Island, near the eastern entrance to Lancaster Sound, the town of Pond Inlet has become a strategic staging and jumping-off point for oil exploration. In earlier times it was a very small community of Inuit hunters, which because of its proximity to one of the great biological oases of the north had strategic value to native survival. Titus Allooloo, an Inuit whose own lifetime has bridged profound cultural change, has served as the mayor of Pond Inlet.

When Allooloo was a child, his family lived entirely from the land and the sea. His earliest memories are of a year-round camp about one hundred kilometres from Pond Inlet, where his family lived in tents during the summer and in a sod house in winter. Two other families lived there also, like his, hunting seals and caribou. There were no "store-bought" foods. When Allooloo was about eight years old, his father made one of his occasional trips to Pond Inlet for supplies. There, he got to talking with a federal administrator, who offered him a job. "At that time, if the white person asked you, it was a command," Allooloo explained. So the family packed up, left their camp, and moved into Pond Inlet. The children were homesick for the old life and the old places, but they had to go to school in Pond Inlet. They did return to the old camp for two months each summer, but then it was back to school in the town.

Later, Allooloo was sent to Churchill for his secondary education, together with hundreds of other Inuit and Dene children. He didn't like Churchill much. The hardest thing to get used to was being scheduled ("programmed") to a strict timetable. The timetable of indigenous peoples has to do with the snow-melt and the tides, the freeze-up and the caribou birthing season, the arrival of the sun and the running of the fishes. It has nothing to do with monocultural schedules. In Pond Inlet at least you could slip away from school at recess to check your fox trap, and it wasn't far to the camp, where the seals were. Churchill was different, and again the boy was homesick. But the first summer when the growing Titus returned to Pond Inlet, he wasn't happy with what he saw. "I suppose it was a change in me, and a change in Pond Inlet, and in the people. . . it wasn't what I expected at all. After that, it was harder for me to adapt—to be happy—in Pond Inlet."

Titus Allooloo is a wise and self-contained man who, although he does not like much of what he has seen, has coped with it. These days he is concerned mostly for his little son. "We don't have a dog team anymore, but there are two dog teams in Pond Inlet, so I take my son out as much as I can, to let him know that there is a land for him. . . to make him understand that there is *freedom*—just out there—waiting for him."

"I don't know how long the old way of life is going to be with us," he says, "but to me it's a unique culture. It's different from southern culture. It helps you keep your sanity, if you go out on the land—see the land, and the animals, and appreciate them. I think I would go crazy if I couldn't get out on the land. So we go out for a couple of days' hunting on the weekends. We don't hunt for sport; we hunt for food, and for the pleasure that goes with it. When you get home from the hunting weekend, your batteries are charged up for the next week. . . . Our psychiatrist is the land: the land keeps our minds straight. If we just stay around Pond Inlet, we will probably start drinking. A lot of what we saw in Churchill wasn't too good. The same thing could happen in Pond Inlet."

What the urban-industrial mind cannot accommodate is that there are cultures in the world, very old ones, that are based not on production and consumption, but on environmental appropriateness. The Inuit happen to be hunters, and that is what their culture is all about. If the people are prevented from hunting, either by the destruction of native wildlife or by their own confinement in monoculture-oriented communities, both their accumulated traditional appropriateness and their individual identities will collapse.

---

24 *Seabirds cannot cope with stresses such as oil pollution.*
25 *Oiled feathers mean death by starvation or exposure.*

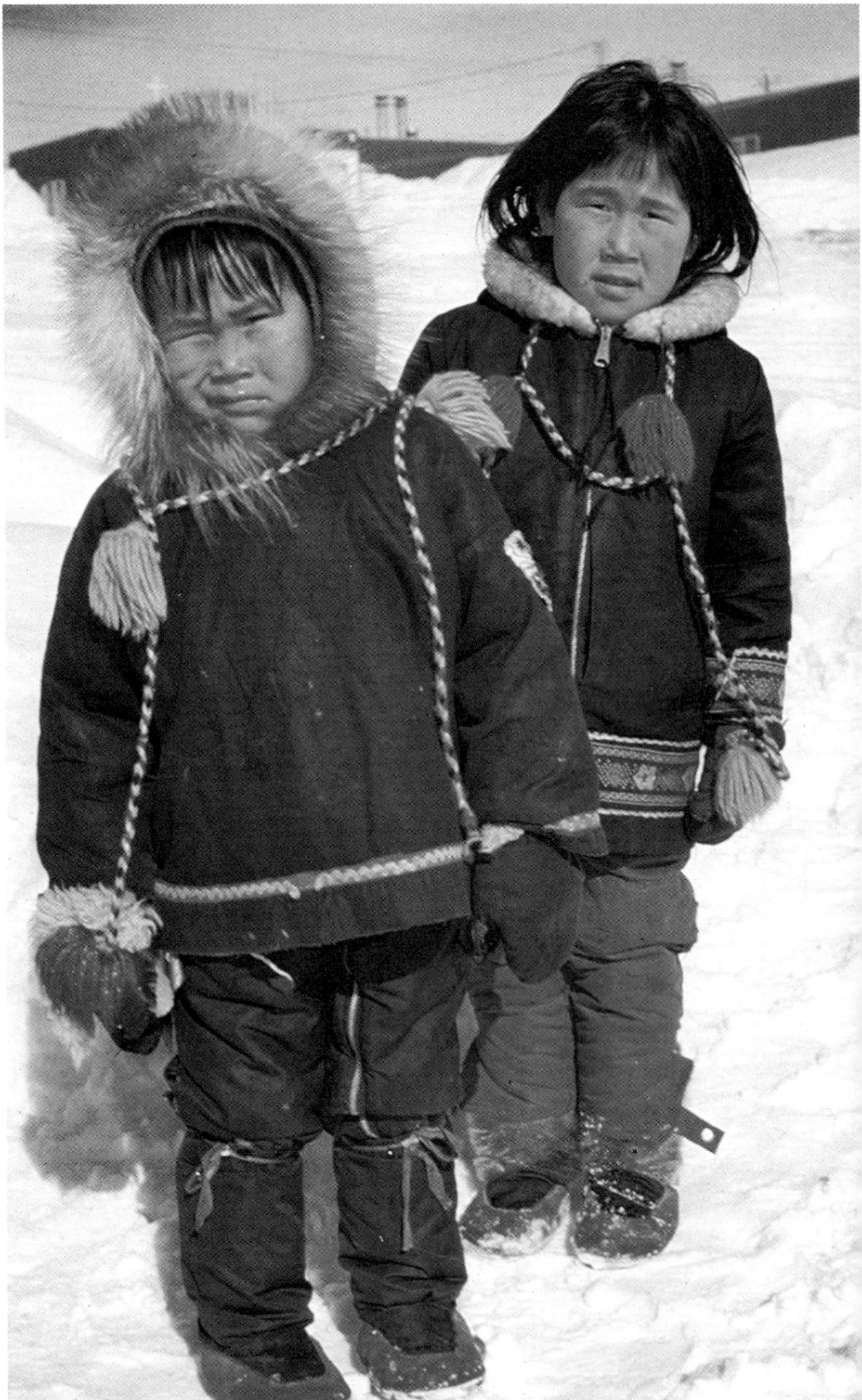

26 *In theory the Inuit still have a choice between cultures.*

27 *Oil clean-up efforts are often primitive.*

28 *(overleaf) Snow geese on summer offshore ice, Bylot Island.*

29

30

29 *Only a gunshot can penetrate the muskoxen's defence circle.*

30 *Barrenground caribou have been known to hybridize with introduced reindeer.*

31

32

33

You cannot have an identity without a social place, and you cannot have a social place without a culture.

The Inuit freely acknowledge that many of the white man's techniques have been advantageous to them. They are not about to abandon firearms and outboard motors at this late date. They too use oil and gas. But they want to make the decisions about how the land is used, because the land is their culture and their identity. They want to see the animals protected. Allooloo voiced the root fear. "Even if they don't destroy the animals, they are going to destroy the minds of the people. . . . They will turn to alcohol."

The Inuit want to make the decisions that will determine their own future, but the long tradition of autocratic paternalism that has flowed from Ottawa and Yellowknife is still very much alive today. The tentacles of the monoculture are long. It may not be the Great White Mother any longer, but the industrial development ethos, as it gathers momentum in the north, displays the same imperial indifference to the subjugation of a culture.

Panarctic Oils have been using Pond Inlet as a logistics site for nearly ten years. Their oil-rig crews have passed in and out of the town on a regular basis for a long time now, and the local people have been accustomed to that; it didn't appear to affect them, or to bother them. But they were quite unprepared for Norlands' entry into the Lancaster Sound region. Norlands had been doing studies for two or three years prior to their 1978 application, one resident reported, "but they hadn't made us aware of the fact that they had been doing them. Once in a while they would land here for fuel, or come down for lunch. But they didn't talk too much, and there were only three or four people around. Nobody really knew what was going on."

Then came the Norlands application to drill in Lancaster Sound. It came suddenly, but the local people responded. Oddly enough, as a local person put it, "no one really expected it to be approved anyway. It gave the people an opportunity to jump on it [at the local hearings], and to have their voices heard as to why they

---

31 *The fortunes of the lemming have dramatic peaks and hollows.*

32 *The arctic hares of Ellesmere gather in large flocks.*

33 *In spring, the arctic fox's coat turns from winter white to summer brown.*

didn't want drilling. I don't think that anybody in the oil industry was upset about [the deferral of a decision]. I think it was used as a sounding-board, almost, as to how the people are going to react to the 'big one' when it comes." The big one has not yet come, but it is coming, and the people are waiting.

At any inquiry into the fate of the Lancaster Sound Region, they are going to want to be heard, fully. Carson Templeton recalls, "I think that at the Berger Commission a fellow by the name of Les Carpenter from Sachs Harbour [on Banks Island] said 'The oil companies come and show us free movies and talk to us in seventy-dollar words, and when we say *no*—we don't want development—they go ahead and do it anyway.' That's the story of the north."

The Inuit are not members of the monoculture; they are different. Their culture cannot be force-fitted to the melting pot. But neither can it be forced to remain in a stereotyped past. Like all societies, Inuit society evolves and changes. As Thomas Berger has said recently, "Native people do not wish to return to the past. They do not wish to be the objects of mere sentimentality. They wish to ensure that their culture can continue to grow and change—in directions they choose for themselves."

The original people want to choose not only the direction of change, but also the timing. In 1977 the Berger report recommended a ten-year moratorium on a Mackenzie Valley pipeline, in order that there might be time for the people to prepare for it. "If it were built now, it would bring limited economic benefits, its social impact would be devastating, and it would frustrate the goals of native claims. Postponement will allow sufficient time for native claims to be settled, and for new programs and new institutions to be established," according to the report.

In the summer of 1980, no more than three years later, a new pipeline proposal for the Mackenzie was put forward by Interprovincial Pipe Line Limited. The local people recalled Berger's recommendation for a ten-year moratorium; they wanted *time*. Then a reporter noticed that the federal government had never acknowledged, in writing, its acceptance of Berger's original recommendations. Phoenix-like, the development imperative rises again.

In the olden days in the north, change was gradual—so gradual as to be almost imperceptible. That is because change took place naturally—appropriately—and natural change is measured in Earth-time, which is the opposite of hurry. Industrial man has forgotten how to wait. In the zeal of its mission, the frantically impatient production machine has no time for externalities, much less one so nebulous and unquantifiable as a culture.

THE CONSERVATION ISSUE

As it has been with the 1977 Berger recommendation for a ten-year moratorium on pipeline-building, so it has been also with his recommendation that "if we are to protect the wilderness, the caribou, birds and other wildlife, we must designate the Northern Yukon, north of the Porcupine River, as a National Wilderness Park," within which the native people would continue to have the right to hunt, fish, and trap. Such a park has still not been designated—or at least there has been no public acknowledgement as to whether Berger's deeply considered advice was ever seriously accepted.

So it has been too with the exhaustive work over a ten-year period of the International Biological Program, which studied and evaluated natural communities all over the world, and made specific recommendations to national governments for the preservation of selected individual sites. In the Canadian arctic, seventy-two unique natural areas were identified as being in need of preservation, six of them in Lancaster Sound. To date the government of Canada has seen fit to take action on exactly one of the seventy-two—and that for a mere two-year period.

The single site that has been awarded this limited recognition is on Bathurst Island to the west of Lancaster Sound, an island that is fairly blanketed by oil and gas exploration permits. Here, for over eleven years the National Museum of Natural Sciences has operated a biological field station under the dedicated leadership of Stuart MacDonald. Its object has been to maintain through constant monitoring a *continuing* record (bear in mind arctic oscillations) of biological events over time. Over that period a growing volume of priceless information has been gathered that is beginning to shed some light on the strange and as yet unfathomed fluctuation that is at the root of arctic vulnerability.

Here, Polar Bear Pass is not only a biological but also a "thermal" oasis. Each spring the ice and snow tend to melt away somewhat earlier than they do in the wider landscape, offering a very slight but still real advantage to the local plants and animals. The sedge meadows support especially rich and varied animal populations: an unusually wide spectrum of birds for these latitudes, as well as caribou, muskoxen and arctic fox. There are polar bears, and a few wolves, and walruses who make use of the marginally earlier ice breakup at the sea edge. Bathurst, especially at the recommended IBP site, is both the beauty and the process of far northern life systems in microcosm.

In spite of this, in spite of the devoted investment of time and diligent care by the National Museum, and in spite of the urgent recommendation of the International Biological Program, the Bathurst site was given a mere two-year "stay" in the face of the insatiable momentum of gas and oil exploration. No other site has been granted even that.

Clearly the limpet-like cemented stubbornness of the federal government and its officials with regard to arctic sanctuaries, wilderness and wildlife reserves, and nature preservation in general, is related to and no doubt springs from the perceived need to conserve and protect development options. There can be no other reason. The arctic is so physically vast that the seventy-two individual IBP sites represent little more, really, than a *sample* of natural areas that because of their individual uniqueness would seem to have a claim upon the conscience of industrial society. But the federal government *will not act*. Its inaction in this matter when seen together with its enthusiastic and zealous participation in industrial advancement in the north brings its priorities into striking and unequivocal perspective.

An additional dimension to this issue, assuming that at least some of the IBP recommendations were to be acted upon, would involve the *protection* of designated nature reserves. Each site, in context of the total area of the arctic, is relatively small, and each would be defenceless against events which might arise as the result

100

of industrial activities in its general neighbourhood. Thus the question of "buffer" zones emerges. We can readily picture a low-lying coastal site which might support a nesting colony of geese, or some precipitous murre cliff overlooking rich arctic waters, or a traditional walrus hauling-out beach as being especially vulnerable. An oil spill even some distance away, but delivered by prevailing currents or winds, could play havoc with such places, even though the places themselves might have been exempted from exploration, production, or use in the logistics of transportation. The effect would be the same as if there had been no formal protection at all. The wildlife would have no other place to go. The question is, then, how large need an arctic sanctuary be, in order to be safe? The answer is, very large indeed. The thinness of the biological envelope in the north and its distribution over enormous areas, together with the seasonal movements of wildlife, mean that protected areas which might seem quite extensive by southern standards might well be insufficient in area to maintain the ecologic integrity of the arctic life system. These turn out to be deeply vexatious issues, and perhaps begin to explain the governmental reluctance to make any preservation commitments whatever.

The complexity and depth of such issues also may begin to explain why no major conservation group or organization has yet identified itself as being flatly opposed to any and all industrial penetration of the far north. All seem to be willing to acknowledge that *some* industrial impact is both inevitable and inescapable, but, with the exception of singling out specific sites such as the IBP recommendations and others, none has said just how much is "some", or just how much is "tolerable". Here we come back to the conundrum involving the threshold of qualitative change. A hard and defensible line simply cannot be drawn. Changes in kind cannot be predicted. The issues become much more than mere vexation; they become more than mere manipulation. They become the deeper metaphysical problems which we shall discuss in Part Three.

101

ENVIRONMENT AND POLITICS

At the outset of this Part we drew a distinction between political and environmental issues. The reader will long since have observed that the distinction is arbitrary, artificial, and quite without meaning. Indeed categorization of this kind becomes worse than useless; it becomes downright misleading. It implies that we can sort out all of our current issues into little neat boxes, and then deal with them in an orderly and systematic manner. Of course it doesn't work that way. If ecology has taught us anything at all it has taught us that nothing—not even an issue—exists in isolation. All things, including issues, are interrelated.

Oil pollution of the sea is clearly an environmental (ecological) issue, but just as clearly the failure of national governments in its abatement is a political one. The absence of oil clean-up technology is a technical issue, but the invasion of the north in knowledge of that absence is a political one. The "one in a million" and the "need to know" issues are obviously both political phenomena which have sweeping implications for the integrity of northern environments. Financing of oil exploration in "frontier" regions is as political an issue as one could imagine, but it allows social and environmental hazards to increase yearly. Environmental impact assessment, so manifestly an ecological concern, turns out, when one examines its practice and procedures, to have profoundly political implications. Social impacts on the indigenous northern peoples are revealed to be not so much political phenomena as they are manifestations of the gradual extirpation and extinction of cultures that evolved out of and in relation to the biophysical environment of the north, and which require the ecological integrity of the north in order to survive. Government reluctance to act for the conservation and preservation of unique natural areas turns out to be deeply related to arctic ecological limitations while at the same time showing itself to be part of the larger political strategy of "development".

Thus it is that our use of labels such as "environmental" or "political" is of little help in our attempt to develop understanding of the nature and the implications of arctic questions. We have been oversimplifying the issues. This is why it is suggested here that although the tip of the iceberg is highly visible in the daily media, to confine ourselves to that stratum of examination is to deny ourselves access to the possibility of deeper comprehension. In the belief that the ultimate resolution of arctic issues may well depend on more general understanding of underlying problems, we shall now attempt to explore the rest of the iceberg.

PART THREE
THE REST OF THE ICEBERG

Arctic hare, Devon Island

105

Subsurface problems are those that seem to give rise to and sustain current issues. The issue of toxic pesticide pollution arises in a society that sees it in the public interest to poison its own living environment; the torture of laboratory animals is justified in the interest of the technology of medicine. Both depend in turn on the fundamental assumption that the advancement of the human good is a more serious matter than any (admittedly regrettable but nonetheless unavoidable) "externality". Similarly, the gradual but inexorable crushing to extinction of the Inuit culture may be seen by many as a lamentable, but inescapable side effect of the more pressing need, which is the fueling of the insatiable urban-industrial monoculture. Some things are simply more pressing than others, and if those latter things have to be called upon to pay the price, that is too bad, but that is the way it is.

Not that the monoculture has cold-bloodedly set out to eradicate the Inuit culture and the wildlife communities of the arctic; of course it hasn't. It is merely a question of the monoculture having certain basic and growing needs, and the satisfaction of those needs being paramount. Since the health and the progress of the monoculture take precedence over all other concerns, *how* it satisfies its needs is a secondary consideration. Any spin-off effect or incidental impact from that process of need-satisfaction thus moves to a tertiary level of importance. We can so begin to arrange the priorities of our society.

It should come as no real surprise, then, that the "need to know" is of more significance than the need to understand arctic biology and ecology. The need to explore for oil and gas takes precedence over the need to know how to clean up afterward. The need to subsidize the extraction, processing and shipment of oil and gas *now* is seen as greater than the need to invest public funds in research for alternative energy technologies. All this is because a society (or an economy) cannot stand still; it must move, must grow, must expand, must "develop". And it must do these things continuously. This is more than a mere imperative. It is an ethos.

In order for a society or a culture to sustain and to nourish a growth ethos or any other governing principle, it is necessary that its members see the world in a way that is compatible with that ethos.

106

The ethos thus dictates the way we see and is itself the product of what we see. (The perception of nature as bloody tooth-and-claw competition is a prime example of this.) A chicken-and-egg relationship is at work here; it is also called "mutual reinforcement". If we see a potato as a "resource" placed on Earth for no other purpose than our consumption, then the act of eating it tends to underscore the position of the potato in a universal hierarchy of relative importance. Just as a lemming both affects and is affected by its environment, so a certain set of ideas both creates a culture and is created and perpetuated by it.

The industrial monoculture perceives the landscape of the world that is not as yet brought into production in much the same way as we perceive the potato. Obviously both the vegetable and the rest of the biophysical community were created for a very specific purpose: to serve the needs of human society. There could be no other reason for their presence on the planet. Further, since human consumption is the reason for their existence, they *must* be consumed. Not to consume them would be to "waste" them, and waste of "resources" is abhorrent to the monocultural orthodoxy. Also, since both potatoes and fossil fuels have a single purpose, for us to deny them their destiny can easily be seen as a crime against both God and nature. The appointed role of the urban-industrial monoculture is to consume.

A critical tenet of the consumption orthodoxy involves *efficiency*. In practice this means consuming something in its entirety, if possible. The polar bear's use of the seal is inefficient in that he may convert no more than 10 per cent of consumed seal energy into polar bear meat, and he even goes so far as to leave waste scraps for the ivory gulls as well. As we have seen, the muskox is somewhat more efficient; he "maximizes" the conservation of energy in a number of ways. On the other hand, the hunting Inuit were clearly primitive and underdeveloped in their minimal, one-dimensional use of the land. Anything less than multiple-use maximization is heretical to monocultural dogma. It follows that the salvation or redemption of both the unfinished Inuit culture and the unused plant and animal communities of the north lies in their unreserved and total dedication to the higher purpose.

Make no mistake: the monoculture *believes* these things. It

107

follows that if there is no readily available alternative to the invasion of the north, then it will be invaded and will be brought down. That may well be regretted by many, perhaps even by a majority, but it will be brought down regardless, because such action will have been *necessary*.

This grim conclusion may take some getting used to, but on the evidence it is inescapable. Perhaps it helps to explain, however, why this book has relegated the popular "issues" to a symptomatic level, and has made a distinction between those superficial symptoms and the much more deeply-embedded causes of the destruction of nature in our time. If indeed there should turn out to be an escape hatch out of the cultural melting-pot and into sane and sensitive environmental appropriateness, that hatch is probably hinged on our better appreciation of the nature of the melting-pot itself, and on our willingness to discover that there actually may be alternatives.

Any choice of highlights, or of points for special examination, is obviously arbitrary. Since everything in any natural community is hooked up to and thus part of everything else, the student of arctic ecology can start anywhere—with a poppy or a clam or a blizzard or a frost boil. If he stays with it long enough, however, whatever point of entry he may have chosen will eventually lead him into wider and wider pathways, toward the presumed but unreachable "whole". Then he would find that an understanding of the whole arctic would in fact be dependent on an understanding of the whole biosphere, which would in its turn necessitate an even wider view. So it doesn't much matter where he starts. The same is true of the exploration of a culture; where you begin is of very little consequence because everything in a complex of ideas plays upon everything else and, as in nature, everything *depends* on everything else.

The following grouping and labelling of problems is for these reasons entirely arbitrary. As with the selection of natural history highlights and sample issues, they are meant to do no more than illustrate the *kinds* of forces that seem to be at work beneath the surface of the arctic "debate", which as we suggested at the outset is not yet a debate at all, but a monologue.

RADICAL CONSERVATISM

To those who study evolutionary biology, one of the most striking characteristics of the process of life change is its inherent *conservatism*. There have of course been very dramatic changes over time, as everyone knows, and one would think that the change from some furry little insectivore of the forest litter into *Homo sapiens* has been anything but conservative. On the other hand, however, the *process* that produced that dramatic change—the process that is called adaptation—seems to be propelled by a force that is as single-mindedly reactionary as that which drives water downhill.

Some years ago the anthropologists Charles F. Hockett and Robert Ascher expressed this phenomenon in what they called "Romer's Rule" after Alfred Romer, the great vertebrate paleontologist of Harvard. The rule goes like this: "The initial survival value of a favourable innovation is conservative, in that it renders possible the maintenance of a traditional way of life in the face of changed circumstances." In other words, the small changes that always turn up in individuals of every generation of living beings, if they should turn out to be advantageous (most of them aren't), will be advantageous in that they allow the possessors of those new attributes to go on doing what their ancestors always did, even though their environment may have changed.

It was in these terms that Romer explained the invasion of the land by the early air-breathing fishes, the famous lobe-finned lungfishes of four hundred million years ago that eventually turned into amphibians, the first vertebrate beings on land. The lungfishes could breathe air after a fashion, and they had stumpy, strong fins. Occasionally, in a period of drought, they would find themselves stranded in a rapidly shrinking pool. At that point, those who could breathe air a little better, and whose fins were a little bigger and stronger, could lurch and struggle and wiggle their way to a safer, deeper pond. Some of these eventually turned into amphibians, some of which much later became reptiles, then birds and mammals. But what the air-breathing and the strange, thick fins *really* allowed them to do was to go right on being the fishes they had always been. Indeed, as their fins grew stronger and their lungs

109

developed it was *easier* to go on being fishes, because it was easier to get back into the water when they had to. The drive was to continue to be fishes; it was conservative. That they changed into something else in the process was in this sense only an incidental result, and that change was only visible by hindsight—much, much later.

This means that adaptive process—innovation—is always trying to maintain the status quo. It is like the constant adjustments that the carpenter or the gunner makes to keep the little air-bubble in the middle of the spirit level. Tilting toward balance. But balance is of course never achievable because the spirit level won't keep still; environments are constantly changing. And just as change is constant, so too the compensating air-bubble adjustments are constant. This is the dynamic of evolutionary adaptation. In the process of being ultra-conservative, living beings change in spite of themselves. In the process of struggling back to the water, the air-breathing lungfishes became amphibians—though they never had intended to do so. A series of insignificant incremental changes over time had resulted in a qualitatively different being—a new *kind* of being. As we saw with both biological communities and with cultures, however, the precise *point* of the "threshold" of qualitative change can never be fixed—merely approximated—and that only long after the fact.

As it turns out, Romer's Rule—the innate conservatism of evolutionary innovation—may be applied quite comfortably to events and contexts that on the surface at least have precious little connection to biological evolution. There can be no better example of the conservative nature of innovation and of adaptation than the contemporary "energy crisis". Leaving aside the fact that no one knows what energy is, and leaving aside also the fact that whatever it is, it streams from the sun constantly and uninterruptedly, and will do so through the lifetime of this solar arrangement, let us look at the "crisis".

The crisis is not one of energy; energy is all-pervasive and continuing, world without end. The crisis is one of oil. Oil is the collected carcasses of innumerable microscopic organisms who died very long ago but who did not decompose in the usual way. Entrapped as they were, probably by sedimentation, the energy in their tiny bodies was not released back to space as rapidly as is normally the case with larger beings. By converting multitudes of these ancient beings into gasoline, we simply speed up their journey a little bit. One of these days we will have unearthed all of them, to rejoin their kindred in the great heat sink in the sky.

But long-dead animalcules are not what the crisis is all about. It is about our millions upon millions of machines that depend on oil. And then again the difficulty is not really our oil-guzzling machines; they were not foisted upon us by the Almighty as a necessary condition for being human. We created the machines, all by ourselves, in a time when there was plenty of oil to feed them. We became as dependent upon them as they were dependent upon oil. In the process, we have become as addicted to oil as our ancestor the lungfish was addicted to water. It follows quite naturally that in our conservative drive to continue to use our oil-fired machines we will go to any lengths whatever either to get oil for them or to devise a substitute for oil which we can use in the same or a similar combustion process. We want to go on being fishes.

We are quite capable, as a society, of committing criminal acts in the search for oil or for the raw material for some synthetic fuel, rather than peaceably looking for other ways of doing things. It is quite possible that in our frantic thrashing about for ways to keep the internal-combustion engine turning we may well, like lungfishes, stumble over some favourable innovation that will allow us to go on doing the same traditional things in the face of changed circumstances such as the absence of oil. Or we may not. We should remember that only a very few of the lungfishes turned into amphibians. Some kept on being lungfishes, but most perished, high and dry. The fossil record is full of them.

It is the (perfectly natural) force of radical conservatism that we now see at work in the arctic. The industrial juggernaut is driven by it. The drive is born not out of greed, nor of aggressiveness, nor even of the lust for corporate or political dominance. It is born of fear. And no fear is so profound as the fear of change.

FEAR OF CHANGE

It may seem paradoxical that the conservative nature of evolutionary innovation is also the source of all of the "success stories" that eventually become new beings. We may be sure, however, that it was fear of the change represented by the quickly drying air that drove the moist-skinned lungfish to squirm his way overland. The old way of life was threatened. Today, the slowly but inexorably diminishing, finite world supply of oil is a threat that is closing in on the industrial monoculture. We can expect it to stop at nothing in its frenzied conservative drive to keep everything just as it is now—or just as it was yesterday. And the momentum of that drive will become so fierce that whatever may have to be "sacrificed" for its satisfaction will be little enough to pay—especially if it is something so remote, uninhabited and presently useless as the arctic.

The possibility of freezing in the dark represents a change few of us would care to experience. The fear of an empty gas tank carries equal foreboding—perhaps more. It seems reasonable to expect that the petroleum industry and its blood brethren in politics and government will not be at pains to dispel those fears, because the greater the anxiety that builds among the public the greater will be the likelihood of the industry being able to further postpone the realization of *its* greatest fear, which is of a change in the structure of economic power. Such a change would represent an upheaval of its environment such as the industrial monoculture has never undergone in its evolutionary career, and might not be able to tolerate.

Everyone knows that it is ridiculously easy for the cynical to work on the conservative fear of change that is in all of us. Demagogues, populists, fundamentalists and parents have always done it. In our time industry and government do it also. And they work hard at it, because if they fail to persuade the public to the brink of paralytic fear, then the industrial way of life—the old way of doing things—could vanish with little trace. It would be a cruel and eviscerating blow to the "development ethic".

The urban-industrial way of doing things depends entirely on growth. Expansion is both its goal and its means; without expansion it must collapse. It is quite literally insatiable; it *cannot* stop.

112

Like the unfortunate who is hooked on drugs, the monoculture must always have *more*, because yesterday's dose is insufficient today. It must manufacture more this year than last, more next year than this. The growth of production depends on the growth of consumption. If the public does not "consume" more fabricated commodities this year than last, and more next year than this, a shuddering paroxysm of fear moves through the entire political-industrial structure. A slowing-down of the economy (which means even a temporary faltering in the public appetite for commodities) is a signal more ominous than any earth tremor, a sign more forbidding, certainly, than any faint far-off stirring of the dogs of war.

But it goes even deeper. To entertain its chronic paranoia the monoculture does not even need a real and present shortening of the public hunger; it fears even the *idea* of change in the escalation of the production-consumption spiral. It fears even the abstract concept of alternatives, because growth is more than a mere desire, more even than a basic need. It is an article of faith and an ideology. The ideology of the urban-industrial monoculture dreads to its bone marrow the merest notion of alternative directions for contemporary society. So "conservative" (in the evolutionary sense) is the drive to maintain the status quo that its defence is mounted not with mere vigour or anger, but with religious fervour.

Incidentally, the reader will understand that none of this has anything whatever to do with "political" ideology in the usual sense. The belief in growth is an industrial, not a political phenomenon. So massive and pervasive is the blind acceptance of the industrial imperative that it is shared virtually without reservation by every conventional ideology in the political spectrum, whether socialist, state capitalist or private capitalist, whether democratic or theocratic, liberal or tyrannical. It is shared across the industrialized world, and thanks to the successful proselytizing of both major power blocs, across the underprivileged would-be-industrialized world as well.

As we have seen, the Inuit and other native peoples of the north also fear change—and with reason. Every day of their lives they can see the implications for their cultures of the insatiability of the urban-industrial machine. Their fear is amply justified; they know

113

their own cultures and they know the monoculture. At least in theory, they still have a choice between them. Their fear is that the choice may be more apparent than real, and that there may no longer be any alternative to absorption.

Fear for the rest of us is grounded in the fact that we can perceive no option at all. Our accumulated tradition has done its work so well that it is impossible for most of us even to conceive of optional ways of doing things, with the result that we fear *any* change.

So it is that the combination of an inherent conservative drive to maintain the status quo with a concomitant drive to grow and expand, will cause our society to do whatever is necessary in order to go on doing the same things it has done since the Industrial Revolution. The industrial expansionist ideology (the "development ethic") will continue its ravenous consumption so long as there is petroleum, or some substitute, to propel it. Every last, usable drop must be extracted; that is what resources are for. Any fundamental alternative would be too frightening to think about.

NORTHERN PERCEPTIONS

Alternative ways of doing things are frightening enough, but alternative *perceptions* of things are downright terrifying. Perceiving goes on inside our individual heads, and we have a very deep, vested interest in what we receive into inventory. It would be bad enough to have to watch things changing right in front of our eyes, but it would be tantamount to self-destruction were we to discover that there actually are ways of seeing fundamental things other than the ways in which we were taught to see them. The missionary is just as afraid to let dancing pagan goblins enter his personal universe as his victim is to part with them. Optional perceptions can be very disquieting, which is part of the reason that most members of a culture tend to perceive the same things in essentially the same ways. Safety in numbers; danger in being different.

It is very easy—too easy—for us to impute to the oil and gas industry a vision of the arctic that is rooted in sheer greed. No doubt there is some of that, perhaps much, but the industrial perception of the arctic rises from other collections of images that are not unique to the oil and gas industry, or to the technocrats, or to anyone else. Such perceptions are the very fabric of our "way of life".

Very few of us are not full-time participants in the advancement of what political scientist William Leiss calls the "imperialism of human needs", or of what Marx described as the "fetishism of commodities". In our society, the vast and overwhelming majority of our needs are expressed in terms of fabricated commodities, for which we look to industry. The needs of industry very quickly become *our* needs. (We don't merely "want" the new microwave oven or the new outdoor barbecue; we *need* them. And industry needs fuel and raw materials in order to deliver the goods.) If industry's need for these things becomes desperate enough, then perhaps we will come around in our perceptions of the arctic. In other words, we'll cave in for the sake of our "way of life", telling ourselves perhaps we can "minimize" the damage to the north after all.

There are many perceptions of the far north; some of us carry optional pictures, but most are deeply ingrained. At a very primitive level there is the arctic "wasteland". Much of the northern land is indeed rocky and apparently barren and empty, which means that it can be seen as a waste because it is not contributing its fair share to industrial human progress. Or, as in the biblical sense, it can be seen as a land which has not as yet been enhanced by the hand of God through the tools of man, and brought into fruitful domestic production. Either way, the land is clearly wasted, and wastelands are not merely frightening, but repugnant. Such lands are "wild"—intractable, uncontrollable, untamed, dangerous. They are wildernesses in the strictest Old Testament sense of the word.

Fear, resentment, and even hatred of wilderness is deep in the cultural heritage of white North Americans, especially those of Puritan or Calvinist origin. To the pioneer, nature was an implacable and unrelenting foe, and that tradition is very much alive today. Its origins lie in the Renaissance marriage of Christianity and science toward the conquest of nature through human reason and technology. The North American varieties of Protestantism adapted it to the pioneer purpose, and later it became the keystone of the edifice

115

that is industrial civilization. Very early on, the conquest of nature was seen not merely as right and proper, but as a fundamental Christian *duty*. The subjugation of nature in the interest of human-kind was a *mission*. It still is.

More recently the focus has narrowed somewhat. The waste-lands of the world (those areas not yet populated by man) have come to be seen not so much as being in the future service of all of humankind, but of *industrialized* humankind. Industry perceives both an absolute right and a solemn duty to "develop" such wastelands, wherever in the world they may be. That is why, as we have pointed out earlier, Inuit uses of the land were and are perceived as incomplete, wasteful, and primitive. If a land is not used by industry, then it is not being used *at all*. This perception of the arctic wilderness is widespread.

All of us at one time or another have heard someone say, perfectly seriously, "Yes, it's striking and awesome, and all that, but—*there's nothing there!*" And in the eye of that individual beholder, there *isn't* anything there. This problem is further com-pounded by the fact that vast stretches of apparent emptiness are in many ways threatening, and thus call for prompt and final subdual. Wilderness wasteland is a deep and ugly affront to the imperialistic mission of industry, and it must be "civilized" together with whatever human (to say nothing of non-human) inhabitants it may have. The higher Christian purpose is, after all, the higher purpose.

Industrialized civilization is assumed to be the natural and necessary (preordained) direction of mankind. It follows that wilderness is even more than a bothersome hindrance. If industrial civilization is the predetermined course of man, then wilderness is *unnatural*; it is an evolutionary anomaly. To be wild is to be the opposite of civilized. To be non-industrialized is to be primitive (at an earlier stage of evolution). To be different is to be foreign, strange, threatening.

Wild, unmolested nature is characterized by diversity, variety, heterogeneity. These, by definition, are the polar opposites to the monoculture. Anything so radically different from industrial civili-zation stands directly in the path of world conquest, which is the ultimate goal of industrial growth. Thus wilderness comes to be perceived as more than a waste, more than an unpleasant and even dangerous anomaly. It becomes the very personification of Amorality—which is the antithesis of progress.

Most adherents of the religion of industrialization deal with their crusade as expeditiously and efficiently as possible in order to get on with the greater good. There are still others among us, however, who cherish the good sporting tradition of honour and respect, even heroism, between well-matched adversaries. For these, wilderness is a *test*—a noble opponent worthy of one's mettle—of whom the vanquishing is its own reward. It must be conquered, like Everest, simply because it is there, in all of its silent austerity, waiting for the ultimate struggle. There is much popular writing and film-making around this theme, and it is also resorted to in many industrial films and other productions. The nobility of the conquered—sometimes a wild river or cataract brought into servitude for hydroelectric power—confers implicit laurels upon the conqueror.

The perceived rightness of the industrial cause is expressed in a multitude of ways. Even in the 1980s the image of "the frontier", in all its romantic trappings, lives. As used in North America (its roots are unmistakeably British), the "frontier" stereotype connotes much more than distant boundaries. It carries a vivid note of challenge. Frontiers are there for one very good and simple reason—for brave and gritty men to bring down, to open, to breach, to vanquish. A frontier is an opportunity for those who have the raw courage and endurance to tackle it. The faint-hearted need not apply. This mystique also persists in our time, albeit more subtly than in the old days, but it is still very much alive.

In contrast, those who see sporting contests, frontiers and crusades as a bit old-fashioned in the space age, and who prefer to demonstrate the human power over nature by way of rational plans, charts, and quantitative analysis, have a perception of the north that is entirely without the primitive raw colouring of emotion. The technocrat coolly and objectively looking upon the frozen north feels no resentment, no greed, no anger, no involvement, no challenge—save to his technique. He perceives the distant wasteland as a "tabula rasa", a blank sheet of drafting paper awaiting its destiny. It is without identity, without purpose, without meaning, until its form and function can be expressed in a design for management. Without a strategy for its future development, it simply does not exist. That which is not rationally planned is nothing.

The diametric opposite to the self-controlled technocrat is the unabashed, totally emotional (and proud of it) person who perceives the arctic as an irreplaceable part of "our natural heritage", which must be preserved for the knowledge, aesthetic appreciation, inspiration and enjoyment of future generations of Canadians. This metaphor is widely used by park planners and conservationists. Thinking of the north as a kind of family heirloom, rather like great-grandmother's diamond ring, to be treasured and appreciated and guarded, but especially to be *possessed* by a series of grateful descendants, has its own implications. First of all, it continues to see nature as a commodity, albeit in a deeply appreciative way, but still as a commodity. Also it somehow overlooks the fact that the Inuit were there first, thousands of years before there were any appreciative Canadians (the Inuit are circumpolar, *northern* people). And, tens or scores of thousands of years ago, when the ancestors of the Inuit arrived, there had already evolved a most beautiful arrangement of living beings. They were not possessed by anyone; they were no one's heritage. They simply belonged.

The perception of white Canadian or even of Inuit *proprietorship* over the arctic is rooted in a system of belief older than the Protestant ethic, older even than the religious mission of nature's conquest through science. The notion of proprietorship draws its nourishment from the immovable assumption that everything that is not human is in the human service; more, that everything that is not human is *owned*—absolutely and without qualification—by the human species. You really cannot go further than that. All of nature is perceived as a human subsidiary.

RESOURCE DEVELOPMENT

It was pointed out earlier that the consumption of nature is more than a mere "given" in our society; it is a mission. Man was placed on Earth not just to inhabit and enjoy it, but to exploit it. The fishes of the sea and the fowl of the air, and the trees and the lakes and the peat moss and the rocks were set in place for a specific purpose: to support the rapid burgeoning and proliferation of the species man. There was no other reason for their existence. They were provided as human "resources".

A "resource" is anything that can be put to human use. Spruce trees are for pulpwood; rivers for hydroelectric power; estuaries and continental shelves for oil drilling; polar bears for rugs and tourist dollars; caribou for meat, skins and trophy racks; landscapes for settlement, habitation and "development". Resources are always *for* something; otherwise they wouldn't exist. And of course what they are for is man, exclusively. It is the concept of "resource" that allows us to perceive nature as our subsidiary.

Furthermore, resources are for industrialized technological man, not even for other races or populations of man. Since the Inuit were not, as it appeared, making the greatest possible use of the lands they had so thinly occupied for so long, the incompleteness of their use must now be corrected, because to do anything less would be a dereliction of duty. High technique must be brought to bear. The technological drive is absolute and unchallengeable; it is so strong, according to theologian and political philosopher George Grant, because "it is carried on by men who still identify what they are doing with the liberation of mankind." It is very easy for such men (who represent majorities in industry, government and bureau- cracy, and who are proliferating like algae in the universities) to perceive the conquest of the arctic not only as a yet unfulfilled human mission, but also as the emancipation of indigenous north- ern peoples from a "primitive" existence. It is even possible for the technocrats to perceive other human beings—not merely natives but their own students and offspring—as "human resources".

The "imperialism of human needs" is in practice today the imperialism of urban-industrial needs. If the Inuit, for example, are not capable of recognizing their own true needs, then it is up to us to educate them. If they cannot perceive the resources that are literally staring them in the face, then it is our mission to enlighten them. First they must be made to appreciate their need of the Christian God and of the English (or French, or Danish) language, to be

119

followed closely by the white man's legal and social systems. In the Canadian north, the priest, the RCMP officer and the trader were more than the purveyors of spiritual, temporal and social rules and regulations. They were the first facilitators in the teaching of urban-industrial needs and imperatives. (It has been pointed out by native people that their own traditional language becomes a badge of stupidity.)

The religious over- and under-tones are profoundly important. Believing as it does in the absolute rightness, propriety and necessity of its cause, the technological monoculture, like the most primitive of contemporary Christian persuasions, sees the world, like the individual, as requiring salvation. It sees all things as perfectible, as requiring "development". All things—landscapes, natural communities, human societies, other "resources"—exist so as to be "developed", which means improved, enhanced, enlarged, perfected, toward the higher purpose. The root concept underlying "resource development" is *improvement*. This is the message that is delivered daily to the world's underprivileged, "underdeveloped" nations. If perceived resources are not improved to their fullest extent, then it is not only a waste to human society, but also man has denied those parts of nature their appointed destiny and thus has fallen short in his unique task. This is the development ethic.

Although the development ethic has obtained a strong foothold and is gathering momentum in the Canadian arctic, it is not yet so advanced there or so firmly entrenched as it has become in tropical and subtropical regions of the world. Such regions have had much longer exposure to the process, and have tended to cast up political leaders who for their own purposes have been quick to accept and implement the ideological package that accompanies "resource development". The process as it has taken place both in colonial times past and more recently in "Third World" ventures gives us a foretaste of what may be expected to happen in the Canadian north.

We have mentioned already the old-fashioned, colonialist use of religion, language and other education. Essentially, the Canadian priest, police officer and trader have done what the first white invaders have always done, which is to systematically break down indigenous cultures, wherever they were to be found. Cultures include traditional ways of doing things, seeing things, believing things, organizing society, and so on. Cultures are environment-specific. If you live in a region where it rains in torrents for part of the year and it never rains at all the rest of the year, or where it is frozen solid for much of the year and the thaw is very brief, or if you depend seasonally on shrimp or fish or rice or seals or deer or whatever, your culture evolves *out of* that way of life and is appropriate to that way of life. The mission of the development ethic, for its advancement, depends on the dismantling and the removal of those indigenous, environmentally appropriate, ways of life.

The development ethic must do these things because it is environment-intolerant. Since its sole purpose is resource consumption and its sole means is also resource consumption, its survival depends entirely on environmental inappropriateness. Environmental "fittingness" would destroy it, because fittingness implies both flexibility and the acceptance of things as they are. Living beings and cultures fit their original environments because they evolve out of and with those environments, not in spite of them. To fit also means to be able to *shift*—to compensate, to flow, to adapt. These things the development machine cannot do. It can do one thing only, and that is to consume. Indigenous ("steady-state") cultures simply have to be removed or assimilated. There is nothing else to do.

The process need not be as brutally aggressive as it was in the days of Montezuma or Sitting Bull, although there are current stories from Brazil. It can be much more subtle in "civilized" hands. One method to which many sociologists have called attention is the presentation by the first white settlers of a "lifestyle model" to which the indigenous people can aspire. This model of course includes fabricated commodities, however worthless, which by their very novelty are attractive, and which, because they are possessed by the conquering or imperial class, are visible symbols of status. Such it was in the north, one day with firearms and bullets, tobacco and rum, later outboard motors, still later nylon, snowmobiles, Coca-Cola and cheesies. Once acquired, such things are hard to give up.

Earlier we touched upon the remarkable adaptability of the human species. We gradually get used to cumulative, subtle differences in things—including ourselves—eventually getting used to qualitative, total change. Native peoples become very used to

121

introduced technologies and commodities, and they even get used to introduced institutions, beliefs, assumptions, and perceptions. The latter are of course part of the larger ideological package which accompanies the technologies and the commodities. None of the contents of the package is so important as what Scott Paradise calls the "vandal ideology"—a position that is totally foreign to the original way of doing things, but fundamental to the advancement of the monoculture. Suddenly the "old ways" are no longer appropriate (the development ethic is environment-intolerant); the total cultural environment has evaporated and the urban-industrial ethos has become indispensable to individual survival.

This, very broadly, is the way in which the imperialism of development is advanced in "underdeveloped" regions. The jungle and the savannah and the tundra are no longer "home"; they are resources. The indigenous peoples, in their service to the new way of seeing and doing things, become alienated from both their physical and cultural environments. The conquest is complete when the original peoples become so dependent upon the new economic system that, far from being able to do without it, they actually *solicit* its advancement. Pressured from without and subverted from within, original cultures thus fling their former identities into the melting-pot of the development ethic.

Sooner or later, in clear realization of the nature and outcome of the process, some elements within the original culture will attempt to identify, revitalize and reclaim their vanishing traditions and other elements of their fast-vanishing identity. Such movements are not unique to North America. The last and final irony is that in order to press their aboriginal case, such groups may finally be forced to do so by adopting foreign (inappropriate) notions such as property rights in a land tenure system that is grotesquely out of keeping with the "old way". The visiting team comes to play in a ballpark they have never seen and by rules which are fundamentally antithetical to the very culture they fight to preserve. This is what author Gregory Bateson called the "double-bind".

Such is the way in which the battle flag of the resource development ethic is carried onward to the heathen. It is small wonder that the conquest is prosecuted without the slightest awareness, much less concern, for the non-human indigenous community. If the only value or purpose of original human beings is assimilation into the expanding monoculture, the fate of non-human beings and of non-living natural phenomena in the service of urban-industrial expansionism cannot be a surprise to anyone. Since the insatiable production-consumption machine must expand merely to stand still, and must expand more in order to grow (it *must* grow), the religion of resource development is its own purpose, its own means, and its own fulfillment.

NATURE'S NET WORTH

If, as it appears to be, the appointed role on Earth of both indigenous peoples and indigenous non-human communities is the service of the urban-industrial enterprise, then arctic "issues" such as those outlined in Part Two can now be seen for what they are: little more than temporary diversionary rear-guard flare-ups, or perhaps "in house" or "all-in-the-family" squabbles which have little impact and even less influence on the trend or direction of the affairs of modern civilized society. Certainly they have nothing to do with the decreed destiny of non-human "resources". All that is at issue, really, is controversy over the means and the timing and the sequence of locations—how and when the arctic community will be brought to its proper sacrificial conclusion. Will the piecemeal attrition we have now continue, toward inevitable eventual qualitative change, or will events in Canada and in other parts of the world accelerate the process? If we assume no change in the modern civilized perception of nature, there is no alternative to a very desolate prognosis.

In the view of the industrialized monoculture, non-human nature has *no intrinsic worth*. It has worth only to the extent that can be demonstrated by a cost/benefit calculus. Obviously if it costs more to get oil out of the arctic (even with huge public subsidy) than it will fetch in the marketplace, then it will not be brought out. That possibility does not seem likely to come to pass. If it ever should come to pass, it will have been because demonstrable, hard calculation has shown that it would be in no one's economic interest—indeed to the general economic detriment.

On the other hand, situations can arise in which the benefit side of the calculation is demonstrable in dollar terms but the cost side is not. Rather than abandon the calculation, we consider the incalculable cost an "externality" which, because it is unmeasurable (qualitative, subjective), cannot be taken into account. Benefit, clearly illuminated in numbers, *can* be taken into account, and is. Qualitative, subjective, unmeasurable incalculables include such items as indigenous cultures, non-human individuals and species, groups of species and associations, landscapes and whole regions uninhabited by man. Such is the fate of the undeveloped wastelands of the world, such as the arctic, when they are put to systematic cost analysis. In their present condition they are worthless and cost nothing; enhanced by modern development, however, they can add up to a calculable benefit.

Such exercises are familiar to everyone. There are however some newer examples of "economic outreach" which, by casting the natural world in the human economic model, allow deeper intrusion by the development ideology. Many contemporary ecologists use economic jargon to justify the conservation of nature. We hear, for example, of the value of the polar bear, not to the biological community but to the "native economy". Dollars for hides. Or we hear of the value of the western arctic snow goose "resource" to the California sport-shooting industry.

We even hear of the "seabird resources" of the high arctic islands—the murres, fulmars, kittiwakes and others—which even though they have no known economic utility of any measure, are with the rest of nature relentlessly commodified. This is no doubt a well-meant attempt to make the birds "respectable"; if they can be seen as "resources" (as having some present or future utility) then at least they can be salvaged from the trash heap of "externalities". Possibly this is a useful short-term tactic, but in essence it is one more reinforcement of the perceived human proprietorship over all of nature.

But economic outreach goes deeper. It has become fashionable to see all of nature as functioning in the same way as a human economy. Nature, we may be sure, knows neither profit nor loss,

124

but "natural economy" is a popular catch-phrase in what has become known as the New Ecology. In the last fifteen or twenty years this point of view has come to dominate studies of biological communities and environments. As defined by historian Donald Worster, the New Ecology "has emphasized the quantitative study of energy flow and 'ecological efficiency' in nature, using the ecosystem idea. Other terms, such as 'producers' and 'consumers' give the New Ecology a distinctly economic cast."

Few widely-used ecology texts today do not dwell at length upon the interpretation of nature as an economic community. In addition to the central importance of the "systems" approach, and the ubiquitous production-consumption metaphor, we find a cost/ benefit machine driven by competition, we find "strategies of ecosystem development", and such economically loaded terms as "stability", "efficiency", "productivity", and "maximization". Even the concept of "niche", to which we have referred, depends for its support upon the prior assumption of a competitive natural marketplace, the currency of which is energy.

The New Ecology, for all of its grotesqueness, is appropriate to its time. Like Newspeak, the language of the technocratic culture, the economic model of nature and natural processes rests well with the contemporary version of the conquest of nature through science and technology, which today consists very largely of the management approach. Like resource development, resource management requires for its advancement no more than the usual vision of man's inalienable right to the rest of nature, except that this time the conquest is achieved not by brute force, but by technique and expertise. Appropriately enough, the old *laissez-faire* approaches to both economics and nature have been replaced by the visible hand of intervention, which may take various forms but always involves prediction and control.

Prediction and control, the watchwords of the techno-managerial class, are thus applied not only to human endeavours but also to the processes of nature, in one stroke both explaining and commodifying non-human phenomena. By bringing nature into the human economic order, it becomes possible to perceive and thus to treat nature as one more extension of the industrial apparatus. As nature, its net worth was zero. As part of the larger system, its worth as a commodity can be demonstrated by economic valuation of its utility. Nature is no longer a stranger to be feared. It just *looks* different.

THE BEDROCK PROBLEM

In this Part we have looked at a few selected examples—or perhaps more correctly, manifestations—of the fundamental problems that appear to underlie our unwillingness or our inability to move in some constructive fashion toward the preservation of nature, in particular the arctic. None of these problems exists on its own, or in isolation from the others; all are interdependent and mutually supporting. All are illustrative of points of view, perspectives or perceptions that are shared, by and large, across our industrialized culture. All exist independently of political, theological or other social ideologies. All are descended from common cultural origins.

The innate and inherent conservatism of societies and of cultures, like that of organisms and communities, arises from a basic and fundamental resistance to change. Fear of change is deep in us, so deep that we may take extraordinary steps to go on doing the same old things—even if it means doing them in slightly different ways. This means, paradoxically, that in order to avoid change, we will and can change. And therein lies, perhaps, a long-term hope, to which we shall return in the last part of this book.

For the present, however, our culture sees the far north and other "undeveloped" regions of the world as an opportunity to go on doing the things to which we have become accustomed. It sees the "resources" of the north as having no meaning or purpose beyond the industrial purpose. That is what they are *for*. Some of us may see the north as a "heritage" for the use and enjoyment of future generations of Canadians, but even here the notion of human proprietorship over nature remains clear and distinct. We continue to view nature as our preserve. We project upon nature the model of a market economy in order that we may "grasp" its processes more surely, both in a conceptual way and as ultimate consumers.

Such problems can be categorized in a number of ways, but since all are interrelated and interdependent, no sorting out can be anything but arbitrary. However there is a theme that runs through most if not all of the contemporary arctic problems and issues, and that is the widespread public expectation of science and technology. Put a different way, this is the pervasive socio-cultural faith in rationality, expertise, and technique. John Kenneth Galbraith coined the term "technostructure" to describe the complex of scientists, engineers, technicians, specialists, experts, lobbyists, managers and executives which "become the guiding intelligence of the business firm." The technostructure in our time is the "commanding power" not only in business and industry but also in government—even, it would seem, in organizations dedicated to the exploration of alternative ways of seeing and doing things, such as universities and research institutions.

Public acceptance and endorsement of the technostructure has too often allowed environmental assessment to become the tool rather than the judge of the resource developer. It is our willing, even fervent embrace of the technostructure that has given birth to the profit-and-loss view of nature—the expectation of a "dividend" from our "investment" in the life process—that characterizes the New Ecology. It has led also directly to the addictive dependence of our society on growth. As Galbraith points out, "profit maximization is not...the central goal of the technostructure. Above a certain profit threshold the members of the technostructure are better rewarded by growth itself." This seems to be painfully true not only of the giants of the petroleum industry but also of the bureaucracy, for both of whom the arctic wasteland represents the last great expansionist frontier.

The public permissiveness of the uncontrolled growth of the technostructure is grounded in our widespread awe and respect for technique, which is the legacy of the religion of reductionist science. By separating and identifying the parts you shall understand the whole, and by understanding the whole you shall bring it under your control. By its very nature (objective, value-free, rational and unemotional), modern science may well be, as Worster says, "an alienating force, always trying to reduce nature to a mechanistic or physio-chemical system with which only an economic relation is conceivable." To objectify nature is probably antithetical to its preservation, but objectification is the life spark and the soul of the technostructure that controls and directs urban-industrial society. Someone observed during the 1980 constitutional conference that Canadians seem to be governed by blueprint and strategy rather than by anything so old-fashioned and subjective as responsiveness. The "rational planning, development and management of our arctic resources" epitomizes this.

But (shall we ever reach the bottom of the iceberg?) our difficulties go still deeper. The techno-managerial ethos cannot exist in a vacuum. It too must have a context. And this is where most if not all of our problems find their source. The several streams of our social, cultural and traditional perceptions of nature and of our relationship to nature are older than Christendom, stronger than nationalism, deeper even than scientific rationalism.

The root problem is *speciesism*. As the Australian philosopher Peter Singer defines it, speciesism is "a prejudice or attitude of bias toward the interests of members of one's own species and against those of members of other species." The pattern is identical, he points out, when the principle of equality is violated by the sexist or by the racist, both of whom, like the speciesist, allow the favouring of his own sex or his own race. The speciesist, which is most of us, favours the human interest over any other interest. This was no doubt understandable in an earlier time, but in the "age of environmentalism" its persistence may seem a little strange.

Just as it is racism that underlies the engulfment of the Inuit culture, it is speciesism that allows the cost accounting of northern "development" to see the destruction of animal and plant communities as an incalculable externality, and that allows scientific members of the technostructure to see living, sensate beings—whether polar bears, murres, or codfish—as "resources". Certainly it is speciesism that permits us to see northern lands and their inhabitants as an enemy, a challenge, a goal, an opportunity, or indeed a heritage. It is racism that commodifies people, and it is speciesism that commodifies nature. Here we find as well the essential rationale for chemical biocides and for vivisection.

Speciesism has its genesis and its evolution in a historic blending of the philosophies, theologies, sciences, and other beliefs and

128

34 *Barrenground caribou crossing Old Crow River, Yukon.*

35 *Male and female king eiders, Bathurst Island.*

36 *Downy young snow geese, Bylot Island.*

37 *Stick-billed murres, Coburg Island.*

38 *Mechanised travel in the contemporary arctic.*

39 *Settlements such as Resolute Bay have replaced the original "wilderness".*

40 *Glacier freshwater run-off on Coburg Island in Lancaster Sound.*

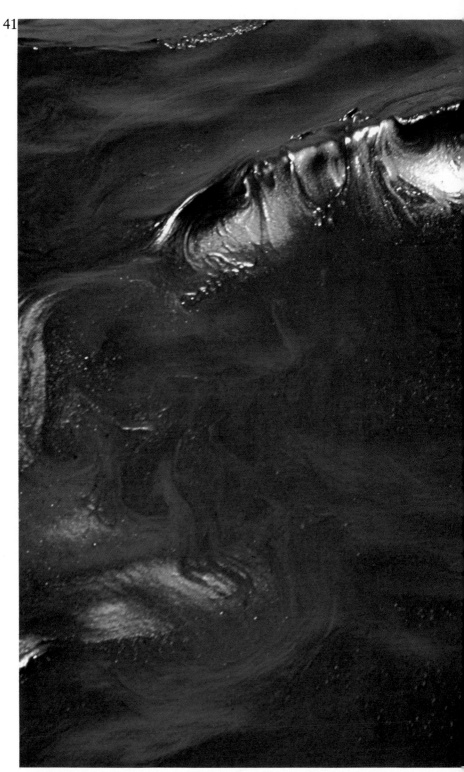

41 IXTOC 1, *an offshore well, poured millions of litres of oil into the Gulf of Mexico.*

traditions that produced the urban-industrial cosmology as we have it today. Through the centuries, the process of cross-pollination, intertwining, combining and recombining of ideas, assumptions and superstitions brought speciesism to its present form. Although it manifests itself in virtually every human endeavour having to do with "resources" or with non-human nature in any way, in none is it quite so highly developed as in the expansionist ethic of the modern technostructure. Here speciesism reaches its nadir, and in it we find the bedrock to which our iceberg of arctic problems is frozen hard and fast.

PART FOUR
ABOUT
SOLUTIONS

Pangnirtung tourist with arctic char.

The intricately close-knit terrestrial and marine communities of the arctic were as long in the making as planet Earth itself. Over immeasurable time, an unimaginably complex series of both physical and biological events swirled and intermingled, streaming together to fashion a unique creation of extraordinary beauty. The beauty of the far north is not merely in its parts or groups of parts, however, not merely in bears and fishes and flowers and seals and lichens and snow geese, not merely in mountains and glaciers and plains and rivers, but also in the fluid meshing of their mutual existences. Stretched to the limit as living beings undeniably are here, there is an awesome delicacy in the elastic tension that holds them all together, in the face of the relentless unpredictability of their environment. But stay together they do, thanks to their innate ability to compensate, their unparalleled knack for appropriate adjustment. Like the walker of the high-wire, their secret is instantaneous responsiveness and hair-trigger timing.

No one knows how long arctic lands and waters were in their primeval natural state before human beings arrived from the Old World. The first northern people appeared a very long time ago, but in the geophysical time scale it was yesterday. Over the short term, however—a term which consists of unknowable human generations—a tentative equilibrium seems to have been reached in some places between the original arctic communities and a relatively small number of human inhabitants. There were few people, both because the arctic cannot support more than a modest population of "top predators", and because the life was demanding and environmental limitations were decisive. But they survived, at an adaptive level of population, and as the human generations turned over they developed a culture—a way of doing things—that was as much the creation of the arctic biophysical environment as of the people themselves. The native people and their culture survived and continued, as we have seen, not in spite of their environment but because their way of conducting their lives did not (at least apparently) significantly damage that environment. It was appropriate, both for the people and for the land.

When white Europeans first looked upon the arctic, they found it dangerous, formidable, hostile. But they also found it promising. The pursuit of the Northwest Passage and of whales and furs and ivory burgeoned from tentative beginnings, and was soon transformed into a strongly mechanized and highly technical quest for non-renewable "resources" in the ground. As the development imperative and all of its appurtenances advanced into the highest latitudes, the original human culture, like the living beings and communities that had long preceded it in time and had supported it through recent history, would have little alternative but to give way before it. The machine had arrived in the garden. What would not be destroyed outright would have to be modified, changed, "adapted", or even assimilated (sometimes consumed) in the interest of the machine's orderly progress. All of this was characterized by a callousness and insensitivity that had long since become the hallmark of the imperialistic mission of the industrial monoculture.

The headlong competition for fossil fuels in the arctic had been under way for a good number of years before there arose any significant questioning of the process among the general public. By the time that criticism of arctic policies and procedures had jelled into a discernible voice, the development imperative had already become solidly entrenched north of 60°, and nothing was to be the same again. Fired now by its own established momentum, the industrial invasion gathered impetus. Very much of this, and most certainly the detail of it, was unknown to the public at large. Both government and the petroleum industry went about their affairs with calm and quiet deliberation, and with little fanfare.

As increasing numbers of individual Canadians and groups began to probe deeper into the conduct of northern affairs, it became clearly evident that there had evolved over time an intimate symbiosis of mutual interest among a succession of governments, their supporting bureaucracies, and the multinational petroleum corporations. Investigators began to report to the public some of the intricacies of the political-industrial web and some of the swiftly growing effects of related field activities on both the natural environments of the north and its indigenous peoples.

The first tangible manifestation of government recognition of the public unease was the Berger Inquiry into the proposed Mackenzie Valley gas pipeline. With his recommendation for a moratorium on pipeline construction to allow sanity a moment to catch up to reality, Berger surprised a great many people, most particularly the high priesthood of the development ethos in Ottawa and in the petroleum industry. He also pleasantly surprised a significant number of citizens who had been wondering when some individual would appear with either the fortitude or the "clout"—preferably both—to apply the brakes to the runaway stampede of industry. That Berger did, at least temporarily, and at least in the northern Yukon. Exploration, drilling and related activities of course continued elsewhere during and after that inquiry. But thanks to Berger's report, the public consciousness of the north had been elevated to an unprecedented level. Knowledge of the plight of both the Inuit and the natural communities had become household property.

Meanwhile, back in the northland, industrial activity further accelerated, but increasingly careful and informed public scrutiny of both the politics and the environmental implications of northern "development" was also growing. Many hot issues emerged; they continue to hold much public attention today. Some of the more critical points have been discussed earlier. Most of them hinge on or are related to the precipitate, damn-the-consequences, full-speed-ahead style that has been so characteristic of the northern enterprise from the outset, quite in spite of the obvious dangers. The enterprise has continued to pick up speed and momentum in spite of a record of horrendous oil spills and blow-outs in various parts of the world and our proven inability either to prevent them or to clean them up afterward; in spite of the known and acknowledged limits to ecologic prediction; in spite of the visible and continuing socio-cultural impact on native peoples; in spite of the obvious weaknesses and the deeper implications of the environmental assessment process. In spite of all of these things, and in the face of the known vulnerability of arctic communities to any disruption from the outside, the conquest has gone forward unabated. Much of its impetus has been maintained by constant reference to the "need to know", and by policies and procedures followed in the extravagant public subsidy of exploration in order to hurry it along, and in the ponderous, agonizing slowness of critical review. Informed critical comment has been either ignored or defused by delay and red tape; research budgets have been grievously reduced. There was not even the time (or perhaps more properly, the inclination) to act on the very modest recommendation of nature sanctuaries put forward by the International Biological Program. There was no time either for so much as a *pro forma* hearing on the potential effects of offshore drilling in the Beaufort Sea.

As we have seen, the origins of such issues are not quite so simple as mere partisan politics, or mere government-corporate "conspiracy", or mere empire-building by the bureaucrats. Although all of these are no doubt important factors, they are underlain by much deeper problems having to do with our perception of the arctic as a commodity, with the sanctity and propriety of the development ethos, and with the unchallenged assumption of the necessary servitude of northern regions to the industrial imperative—all springing from the root speciesist bias of Western man.

To an outside, unprejudiced observer, one of the most prominent and noteworthy characteristics of our society would be its unswervingly childlike demand for and expectation of solutions. Every issue has its answer, every problem its resolution. It is simply a matter of working things through. It may demand a lot of time, and a lot of expertise, commitment, and money, but somewhere there *is* a solution. Thus it is that at the conclusion of any lecture or essay or book that sees fit to concentrate on the identification and description of issues and/or problems, the final "wrap-up" is expected to consist of the presentation of solutions. Very well, it is said, you have laid out the question; what is the answer? What do we do now? It is interesting that when you reply that you don't know the answer, it is concluded that you don't understand your own question, or have got it wrong. Without a good, hard, three-dimensional answer, questioning is meaningless. Let's stick to things that have solutions, in the here and now and by conventional means. Let's not rock the boat.

There is some reason to think that in our critical exploration of public affairs it is this "bottom line" solution-fixation that tempts us to remain with the "issues" and not to investigate more deeply. The issues can, after all, be resolved. Clearly there *are* solutions to, for example, our gratuitous subsidy of industrial financing in the north, to the presently uncontrolled advance of exploration, to the weaknesses of environmental assessment, to the under-funding of basic research both in arctic biology and in alternative energy technology. These have relatively simple answers. The hindrance to their resolution is only political, and that could be dealt with, assuming a sufficiently evolved public will. It would be difficult and no doubt painful, but it could be done.

The "politics" of northern affairs have been referred to throughout, and will not be repeated here. Thanks to the public information roles that have been assumed by such groups as the Canadian Arctic Resources Committee and others, by an increasingly alert, insightful, and demanding press, and by many other individuals, all of us out here in the public have continuing access to "political" aspects of the arctic oil question. If we do not take advantage of information that is freely, abundantly available to us, then it is not the fault of those who dig and ferret it out so relentlessly and who interpret it for our understanding on virtually a daily basis. We do not suffer nowadays for want of information, but perhaps we do suffer for want of individual and collective will. We have no excuse whatever.

The remedy that is most commonly offered for arctic ills would be a decision (whether "political", in the conventional sense, or not) to immediately upgrade both the volume and the quality of basic northern scientific research, toward an information base that would allow rational, long-term planning and decision-making. This would be the only sane and realistic way to go. Clearly such research must precede, not follow, intensification of industrial activity in the north. A temporary "stay" of industrial penetration in the meantime would allow us to better understand what we are getting into, and that greater understanding would enable us to decide intelligently what areas should be off or on limits, what ways of doing things would be safest (or least damaging), what procedures should be followed in the best interests of the native people, and so on. Let us—perhaps for the first time in the history of industrial development—look before we leap.

Such an approach would be eminently reasonable, and entirely "realistic" in the most contemporary sense. It would acknowledge that the industrial penetration in fact exists and will in fact continue; it would insist that from this point on, the process would be less headlong and more considered, less fragmented and better managed, less autonomous and better controlled. Its goal would be the long term, not the short, and the means would be appropriate to the

136

peculiar environmental and social conditions of the arctic. Obviously if the public were to force such an approach upon government and industry, and not take "no" for an answer, most of the current issues could be brought to rein, or at least reduced to manageable proportions.

On the other hand, this "solution" does not of course address the problems, and it does not do anything whatever about the insidious process of destruction by insignificant increments. Indeed it would likely mask even further that subtle attrition simply by conveying the impression that everything is now under enlightened management. This would be a certain way of intensifying its ultimate impact.

Rational planning and decision-making are the basic tools-in-trade of the development ethic as we have described it here. Resource management is understood to be simply the modern technocratic approach to the advancement of the same industrial cause that has been with us for generations; all that have changed are the trappings and the style. The goal is identical; only the means have been refined. To plan the orderly development of the north is to take industrial activity as given, and to make its way easier in much the same fashion as the environmental assessment review process can make the way smoother for the developer. If this approach should come to pass, nothing will have been altered but the rhetoric.

The only plan or policy that will save the arctic is the exclusion of industrial man. Industrial man is already there, however, and is not about to withdraw peacefully. It would thus behoove us, perhaps, to suspend our visions of "saving the arctic" from industry and to concentrate on rescuing what is left of it. Here let us make the assumption that we are bent on saving what is left of it not merely for our sakes but for *its* sake. (There is a world of difference, but for purposes of argument let us assume that such a shift in our position might be possible.) How would we go about stemming the tide of further destruction *now*?

Clearly there should be an immediate total freeze on industrial activity—a Berger-style moratorium, but in this case an open-ended one that would not be lifted until there was incontrovertible proof in hand of the certain consequences of further exploratory activity.

137

This would be an enormous burden of proof for government to satisfy; the public would be required to judge how adequately government fulfilled this duty. All of this would cost a great deal of money, no doubt, much of which we would probably make up at the gas pump, but it would also save a great deal in the way of corporate subsidies.

Once we had stalled the machine and locked up its ignition key in some safe place, it would be possible to begin a careful inventory of the situation. We would not launch another of those exercises so beloved by the "comprehensive" planners, a "capability" inventory, but rather an inventory of the *damage already inflicted*. It would not be necessary to survey those areas not yet penetrated; indeed it would be essential to avoid them. The only inventory we need of the arctic is of the existing record of our accomplishments. A map of *existing* townsites, drillsites, campsites, and exploration and drilling permits tells us all we need to know—already. If we were to superimpose on such a map the locations of proposed extracting and processing sites, pipelines and marine transport facilities, we would begin to appreciate the implications of incremental change. It would be relatively easy to extrapolate from existing installations to the "larger picture".

Having done our inventory we would then apply the principle of *triage*, a term used by military medical services. At a field hospital following heavy action, the wounded are divided into three groups: those who cannot survive in spite of any treatment they might be given; those who will survive without any treatment, in spite of pain; and those who can be saved if they receive care immediately. The third group is the first and perhaps the only one to receive attention, depending upon the availability of supplies and medical personnel.

As William and Paul Paddock describe *triage*, "It is a terrible chore for the doctors to classify the helpless wounded in this fashion, but it is the only way to save the maximum number of lives. To spend time with the less seriously wounded or with the dying would mean that many of those who might have lived will die. It would be a misuse of the available medical help." It seems shockingly cold-blooded, but there are times when there simply isn't the

138

staff to cope with great numbers of wounded, and the hardest of choices have to be made.

The Paddocks used the *triage* thesis in their study and discussion of world human overpopulation, famine, and Third World aid. Such an approach to the Canadian north, where hard decisions are also in store, is worth considering. There would emerge from such an evaluation of our inventory a list of places which are total write-offs; there would be others which could recover over time, simply by being left alone; and there would be those which would require careful rehabilitation and manipulation in order to be able to recover. This is of course an entirely utopian recipe in view of the relatively primitive knowledge of arctic ecology, but the very process of preparing for rehabilitation would necessarily involve an immediate upgrading and intensification of basic research.

To recommend that such research be confined to areas already degraded or at least disturbed is to invite the comment that research "controls" are needed; that is, pristine areas which have never been "impacted upon" and which therefore serve for comparative purposes. Perhaps this is best answered by a further reference to medical affairs; we don't run experiments on healthy people. Neither should we run experiments on healthy biological communities. Let us study the anomalies, not disturb existing well-being. It is harder and more time-consuming, but as in medicine, it is more humane. When we do experiment on healthy people, we use volunteers, after having obtained their signed consent. There are no volunteers in the arctic.

After the *triage* evaluation, we would be in a position to isolate those areas for which treatment is or is not indicated. At this stage we would apply the principle of *enclosure*, which is the usual approach to the planning of parks and nature reserves in the south. We are accustomed to "setting aside" natural areas for their protection, or for our recreation or study, by means of putting fences around them and letting the process of urban-industrial development continue outside and around them. Industrial man is excluded; nature is enclosed. Such enclosures, depending upon their size, shape, proximity to each other, intrinsic peculiarities and other factors, may or may not have the internal strength, richness,

139

resilience or adaptability to survive over the long term. They are islands in a monocultural sea.

In the north, we would have the opposite of this. The fences would enclose industrial man, not nature. Outside the fences, the sea of heterogeneity would be permitted to continue in its own way. The flow of life would continue around and past these industrial caissons, within which technological man could continue his works on sites already heavily or irretrievably changed. As explained in Part Two, the sheer physical vastness that is required for healthy, unimpeded life processes in the arctic indicates a completely different approach to the establishment of defended areas from the "island" technique we use in the south. But since the arctic is wholly unlike any other living community, the application of any form of conventional wisdom must be considered risky.

Unfortunately, this scenario has formidable limitations over and above the very practical question of the "realism" of an industrial moratorium. Who, for example, is to decide on criteria for the evaluation of areas, and on the basis of what knowledge? Having evaluated an area and declared it a candidate for rehabilitation, what would one do next? How? Who would decide, and how, on the ultimate size of the "industrial parks"? And so on.

Also, the mere sorting out of already degraded areas and their establishment as industrial enclosures would completely ignore the dangers of transportation. Even assuming that oil and gas extraction facilities were to be fenced, and that the further spreading of exploration were to be curtailed, we would still have the transportation problem. The shipment of oil and gas may well turn out to be the most dangerous aspect of the entire enterprise. A new proposal from a group of companies that includes Petro-Canada would have gas extracted on Melville Island at Drake Point (site of the 1969 wildcat blow-out) shipped by pipeline to a liquefication plant, thence by tanker through Lancaster Sound, Baffin Bay and Davis Strait down past Labrador and Newfoundland to the Maritimes and the St. Lawrence River. Dome Petroleum's oil from beneath the ice and sea-bottom of the Beaufort will require either the resurrection of the Mackenzie Valley pipeline or a tanker fleet to move a mind-boggling 100,000 barrels of oil per day by 1985 and a staggering 1.2 million barrels per day by 1990. It is estimated that the 1985 output would require two tankers, the 1990 volume a fleet of twenty-four! We can expect comparable projections from the bringing on stream of any and all existing discoveries throughout both the mainland and the high arctic archipelago. (Remember that the exploration would not be going on now were there not at least the implicit understanding that what is found will be marketed. That is what the "need to know" is all about.) In other words, the logistics of industrial support and *year-round* transportation through the ice may be of much greater significance than what might in future be happening in any enclosed industrial "park". Such realizations tend to cruelly knock the props out from under any proposal that would permit any industrial activity to continue at all. Our *triage* suggestion, in the face of the transportation question, could thus be seen as little more than a theoretical exercise.

Obviously what *triage* is all about is the cutting of losses. It accepts the reality of substantial casualties, and does what is possible to make the best of what is left. If Canada were to adopt such an approach to northern policy, it would be the same as acknowledging that at least some arctic lands and waters are expendable. It would define those lands and waters as "externalities". To make such an admission would be to prepare the way for the inevitability of total collapse. *Destruction by insignificant increments is the inescapable result of all development-oriented land-use planning that relies on systems of classification.* The process goes on all around us in more southern regions, all the time, resulting in the inevitable breakdown and collapse of even rich and resilient temperate-zone communities. For reasons we have seen, the arctic cannot tolerate even nibbling at its edges, much less incursions into its vitals. The high-wire artist has enough to contend with; he could not survive so much as a pat on the back.

If all this is indeed the case, then what would be a "realistic" response to industrial development pressure? It may be that there is none—or at least none that can be accommodated within the development ethic. It may be that the solution is not one of means, but of ends. It may be that the answer is to be found in what we want the north country to be like after the due passage of our own time. We shall not find the answer to that in strategic management studies. We can find it only in our hearts.

Very simply, the arctic question is a moral and ethical one. To cast it in any other terms is to ignore the nature and the quality of what is happening. When an individual human being is indifferent to the nature and quality of his acts, and of their effects, we call him a psychopath.

From time to time we hear of the "rape" of the arctic by industrial man. This is a good strong word, and it makes us listen. But it is very misleading. Rape is the forcible violation of an individual's right to give or to withold consent to an act; it is the transgression of an individual's freedom of choice. To apply the notion of rape to the industrial invasion of the arctic is to imply that the biological communities of the arctic and their constituent members have some structure of rights that is being impinged upon. But nature has no rights in respect to monocultural man. The arctic has no say in this, no freedom of choice, no right to grant or to withhold consent, and no protection in any statute book. The only protection the arctic has, like any potential rape victim before the fact, is that which might be provided by some lingering spark of decency and compassion in the strong and powerful. And it has no later recourse; at least a violated human being can bring a complaint after the fact. The notion of rape as applied to any aspect of non-human nature is, like so much of the arctic rhetoric, absurd.

It was pointed out at the beginning of this book that the bewildering dust storm of interacting arctic "issues" and conflicting charges and countercharges may be seen as absurd because it lacks genuine coherence and context; there is only one protagonist and it is us. There is no debate whatever about ends; the issues we have discussed are entirely matters of detail, of means. In actuality the debate is an in-house policy discussion on how to go on doing the things we do now. The debate is about how to go on being lungfishes.

The ultimate challenge to the adaptive resilience of the lungfish was a relatively sudden and quite dramatic change in his environment—a drought. This was no mere threat; it was a total emergency. The current change in the environment of the industrial monoculture is, by comparison with that of the lungfish, paltry. Most of the environmental perturbations experienced by industry at the present consist of advice on how to clean up its act, or how to develop the first faint stirrings of an environmental conscience. The profit-making machinery might have to give a little in order to beef up environmental safeguards, or in order to fund research, or in order to pay to the public treasury a modest tax levy. It might have to forgo bumper profits this year in the interest of a predictable future. But none of these things can by the wildest leap of imagination be interpreted as a "threat". They are no more than matters of fine-tuning; the "thrust" of the development juggernaut is completely unchanged by vicissitudes such as these.

But a more profound change is obviously coming, and that will be the depletion of petroleum in the ground. This moves us much closer to the plight of the stranded lungfish; the monoculture must have fossil fuel or perish. Of course, in his time, the lungfish believed with equal urgency and conviction that he had to have water; as it turned out, he didn't. His requirements changed, and he survived. But he wasn't the old lungfish any more. He was something new—something that the world had never seen before.

In its forthcoming paroxysms, the monoculture will fight against change as valiantly and stubbornly as any grounded lungfish. Since there are two ways of becoming extinct (either disappear entirely or change into something new and more appropriate), the future for Western industrial society seems brilliantly clear. In order not to change, we will change. And since this is going to happen whether we like it or not, it would seem to be the part of intelligence to review the options. It would also seem reasonable to apply ourselves rather less to fine-tuning and minor fiddling adjustments (the "issues"), and somewhat more to anticipation of change *in kind*. From qualitative change there is no escape; the best preparation is to exercise our remaining flexibility.

The lungfish could not know that his very fishiness was soon going to be inappropriate to the point of liability, and that he would eventually relinquish it for his own good. Our society takes a great deal of pride in its capacity to cast alternative futures, to weigh options, to manage its affairs, and to take wise and considered action against tomorrow's possibilities. All we really do, of course, is react. And there is nothing wrong with that; it bailed out the lungfish. It is still puzzling, however, that in full knowledge of the rapid decline in the available supply of oil, we continue to behave as though the shortage didn't exist. Or, we steadfastly deny *any* shortage, alleging that the industry, for its own reasons, promotes the idea of one (the industry figures are, after all, all that we or the government have to go on). Or, we change our minds again, acknowledge a shortage, and in the same breath behave as though at the penultimate moment, something will come to our rescue out of the blue. Or, while admitting the finiteness of the supply, we scramble to extract the last barrel while pretending that more will be discovered. No doubt it will, but it too will be depleted, and a very large price will have been paid for it.

Our responses to date have consisted largely of discussion of the reasons for the dwindling oil supply, and of ways of increasing it. We turn to intercontinental diplomacy, to the refinement of exploration technology, to the penetration of new "frontiers", to greater capital investment, to anything, it seems, except acceptance of the need for alternative means of fueling the apparatus of industrial production, or—the unthinkable—the need for alternatives to massive-scale industrial production itself. No effort is too great, no cost too high to pay, in the effort to delay the hard decision. Rather than face and admit the presence of impending change, we run around drilling holes in the ground. We cannot stop because we know we cannot handle the change. We refuse to admit that there can be no technical answer to a moral problem.

To the life community of the arctic, it matters not at all for what aspect of our behaviour it must pay the price, only that it must pay it. Whatever the reason—political-bureaucratic ambition, multi-national ravenousness, the failure of high technology, the blindness and insensitivity of the industrial monoculture, the fanaticism of the development ethic, or even centuries of intra-Semitic hatred—the arctic is being called upon to subsidize the final jerking spasms of an inappropriate and maladaptive evolutionary aberration, the modern urban-industrial monoculture.

In view of the indisputable fact that there *are* alternative options to the destruction of the arctic for the sake of a very few more years' worth of oil, the inexorable process of creeping attrition is all the more dismaying to observe. It is as though those who press the conquest of the north, from the roughnecks on the ice to the decision-makers in the corridors and boardrooms were not sensate, thinking human beings at all, but automata in the mindless service of some force far greater than the accumulated wisdom of mankind, some intent more powerful than all of human goodness.

The ultimate question seems to be that of timing: do we bring down the arctic now, or later? To say that we never will, or that if we do, it can never return, would be to speak too quickly, because "never" is a very long time. But this much we do know: if we do bring the arctic down, either now or later, then somewhere in this or some other solar system a cloud of cosmic dust must coalesce, must fuse, and must spawn new planets. One of those new planets must be biologically habitable, and four thousand million years of evolution must ensue, before the like of this living community can come again.

Whether this matters to us or not is really what the arctic question is all about. Given the frightening time dimensions of evolution, we might excuse ourselves for not thinking in terms of biological processes, let alone geophysical events. But, on the other hand, given the unreserved awe and reverence in which high civilization is held by most of us, the deeds we do in the furtherance of that civilization may seem to some of us paradoxical, at the very least.

To an unprejudiced observer from some other place, some other time, or some other planet, the deeds we do in the advancement of high civilization might appear somewhat contradictory, perhaps even bizarre. He might look into the "values" we so self-righteously profess, or into the "ethics" that fill our libraries. Having judged our moral development from the point of view not only of our protestations but also of our actions, he might then run a cost/ benefit analysis of the evolutionary appropriateness of the civilized human mind. Having come to certain conclusions about both our morality and our intelligence, he might then turn to examination of our sanity.

Satisfied as to that point, and turning to leave, he might notice a slowly melting iceberg, stranded high on a pebbly arctic beach. Funny, he says to himself, last time I noticed that, it was still in the water, frozen fast to the bottom. It seems to have broken loose . . . the tide is falling, and it's getting warmer. . . . Times seem to change here, too . . .

SELECTED REFERENCES

The following references, some of which are mentioned in the text, are meant to represent no more than an appropriate sampling of the kinds of literature that bear on questions such as the industrialization of the Canadian arctic. Some are background texts on physical and biological aspects of the north; others deal with identified political and/or environmental issues. Still others, while not necessarily directed to northern affairs, examine problems having to do with, or the social and cultural roots of, the development ethic and the industrial imperative. There are even those that offer alternative directions for contemporary society. Each title listed here will direct the reader to many more; the wider implications of "arctic oil" are as vast and complex as our accumulated cultural heritage itself.

Banfield, A.W.F. *The Mammals of Canada*. Toronto: University of Toronto Press, 1974.
> The basic reference text; especially useful for distributions, descriptions, summaries of habits. Selected references. Illustrated. The author has wide personal experience in the arctic.

Barry, T.W. "Observations on natural mortality and native use of eider ducks along the Beaufort Sea coast." *Canadian Field Naturalist* 82(1968):140–144.
> An example of the critical timing relationship between seasonal activities of wildlife and the behaviour of ice.

Bateson, Gregory. *Steps to an Ecology of Mind*. New York: Ballantine Books, 1972.
> A collection of brilliant essays in the "ecology of ideas" by one of the genuinely interdisciplinary thinkers of our time, with special reference to alternative "realities", concepts, and perceptions. Relevant to the deeper strata of the iceberg.

Berger, Thomas R. *Northern Frontier, Northern Homeland*. Toronto: James Lorimer & Co., 1977.
> The report (2 vols.) of the Mackenzie Valley Pipeline Inquiry. Both an historic document and a simple yet comprehensive account of what happened, of the evidence, and of the reasoning behind Berger's decision. Essential background to northern affairs, policies and procedures, and in its sensitivity to and understanding of northern environments and indigenous peoples, a benchmark.

_____. "A Glance at History." *Northern Perspectives*, vol. 7, no. 4, 1979.
> Excerpts from an address to the Canadian Ethnology Society, February 1979, published in a special issue of *Northern Perspectives*, "Native Rights in the New World". Native problems in the north in context of the history of white-native relations in the New World since the Spanish conquest.

147

Bisset, Ronald. "Quantification, Decision-Making and Environmental Impact Assessment in the United Kingdom." *Journal of Environmental Management*, vol. 7, 1978.
An important and disturbing comment on both method and style in environmental impact assessment, with special reference to the preparation and presentation of materials for review.

Britton, Max E., ed. *Alaskan Arctic Tundra.* Washington: Arctic Institute of North America Technical paper no. 25, 1973.
Proceedings of the 25th Anniversary Celebration of the U.S. Naval Arctic Research Laboratory. Technical papers on arctic terrestrial research, including permafrost, North Slope morphology, soils, microclimates, limnology, tundra vegetation and lemming ecology. Technical editor Wade W. Gunn.

Bruemmer, Fred. *Encounters with Arctic Animals.* Toronto: McGraw Hill-Ryerson, 1972.
Bruemmer's magnificent photographs with a deeply knowledgeable and compelling text. Natural history is deftly interwoven with the plea for preservation.

Canadian Arctic Resources Committee. "A Submission to the Federal Environmental Assessment and Review Process Hearings on Exploratory Drilling by Norlands Petroleums Limited in the Lancaster Sound Region." Ottawa: CARC, 1978.
An exemplary "intervention" in the EARP hearings which painstakingly reveals many more dimensions of the relevant issues than "normal" procedures usually manage to do. This single document is a substantial education in the issues.

_____. *Mackenzie Delta: Priorities and Alternatives.* Ottawa: CARC, 1976.
Proceedings of the Ottawa conference of December 1975. Includes technical and political papers, and editorial remarks, with contributions and lessons from the U.S. and Norway.

_____. *Northern Perspectives.* Periodic newsletter on northern affairs published at 46 Elgin Street, Ottawa, Ontario, K1P 5K6. Subscriptions free of charge from the CARC office; back issues sixty cents each. Absolutely essential as continuing background, with analysis and interpretation of the issues.

Canadian Wildlife Service. *Report Series.* A series of occasional publications on wildlife subjects, many of which have had to do with arctic and subarctic species of birds and mammals. These are basic scientific contributions, well referenced, often well illustrated. Formerly under the banner of the Department of Indian Affairs and Northern Development, now Environment Canada.

Devall, William. "The Deep Ecology Movement." *Natural Resources Journal*, vol. 20, no. 2 (April 1980).
A landmark discussion of the two "streams of environmentalism", or strata of ecology. "Shallow" ecology, as defined, deals with issues (pesticide pollution). "Deep" ecology is concerned with the root value system that

148

condones pesticides in the human interest. A work of scholarship and insight for those who choose to explore the rest of the iceberg.

Dosman, Edgar J. *The National Interest: The Politics of Northern Development 1968–75.* Toronto: McClelland and Stewart, 1975.
A brilliant and disquieting analysis of the national and international politics of northern affairs during the critical period when the industrial invasion of the arctic gathered its greatest momentum. Required for understanding of the nature and process of "high level" decision-making in the bureaucratic context.

Dunbar, M.J. "Keynote Address", in *Marine Transportation and High Arctic Development: Policy Framework and Priorities.* Ottawa: CARC, 1979.
The publication is the proceedings of a CARC-sponsored symposium in March 1979, of which Dunbar was chairman. His opening remarks centred on the present state of basic arctic research. The conference addressed many aspects of science policy, social and environmental policy, regulatory problems, and international aspects of arctic operations. Valuable background to the (perhaps insoluble) dilemma of transportation policy in high latitudes.

Ehrenfeld, David. *The Arrogance of Humanism.* New York: Oxford University Press, 1978.
A distinguished scientist moves from the "issues" of biological conservation, in which he has long been involved, to the "problems"—the hidden assumptions and beliefs, the institutionalized perceptions of human power and control which underlie the humanistic traditions of our civilization, and in which are portents for all of nature — including us. Essential to fuller understanding of all of the iceberg.

Evernden, Neil. "Beyond Ecology: Self, Place, and the Pathetic Fallacy." *North American Review*, vol. 263, no. 4 (Winter 1978).
How ecology, in aspiring to become a conventional science, has forgotten the aesthetic dimensions of place and belonging, which are what ecology is really all about. Environmentalism involves the perception of values, and values are the coin not of science but of the arts. Environmentalism without aesthetics is no more than regional planning. More about the rest of the iceberg.

Galbraith, John Kenneth. *The New Industrial State.* Boston: Houghton Mifflin, 1971.

_____. *Economics and the Public Purpose.* Boston: Houghton Mifflin, 1973.
Household word though the name of Galbraith may be, conservationists would do well to draw on his insightful and original analyses of public and private economies, the giant multinational corporations, the "technostructure", and the enormity of the corporate planning system. Sheds implacable light on the economics of the development imperative, which entirely permeates our society.

Gamble, D.J. "Destruction by Insignificant Increments." *Northern Perspectives*, vol. 7, no. 6, 1979.
An appeal for "the kind of long-term research and planning mechanisms that will make rational decision-making possible" in circumpolar oil and gas exploration and development. A straightforward and well-argued case for the planning approach to arctic offshore industrial activity. It is accompanied in the same important issue of *Northern Perspectives* by "An environmental research and management strategy for the eastern region": a discussion by Ian G. Stirling of the Canadian Wildlife Service, Ron R. Wallace of Dominion Ecological Consulting Ltd., and Gerry T. Glazier of Petro-Canada. In its entirety, this issue of *Northern Perspectives* provides invaluable insights in its encapsulation of the planning approach to northern affairs, as discussed in Part Four of this book.

Godfrey, W. Earl. *The Birds of Canada*. Ottawa: National Museum of Canada, 1966.
The basic reference. Companion to Banfield's *Mammals*. Valuable distribution maps, data on field marks, habitat, nesting, ranges. Illustrated.

Godlovitch, Stanley and Roslind, and John Harris, eds. *Animals, Men, and Morals*. London: Victor Gollancz, 1971.
An enquiry into the maltreatment of non-humans, principally from the philosophic point of view. This collection of essays looks first at factory farming, furs and cosmetics, animal experimentation and recreational killing; then at the moral framework and the sociological perspective. A vivid revelation of "speciesism" in practice and a rigorous discussion of the "moral and logical muddles" into which our basic assumptions about ourselves lead us.

Gordon, Walter. "Commentary", in *Mackenzie Delta: Priorities and Alternatives*. Ottawa: CARC, 1976.
Pithy and cogent remarks by the former Minister of Finance at the 1975 CARC conference. Special emphasis on the multinational petroleum corporations and the problem of foreign control.

Grant, George. *Technology and Empire*. Toronto: House of Anansi, 1969.
Perspectives on North America from a distinguished philosopher and theologian. A collection of magnificently troubling essays in political philosophy which bear directly on the fundamentals of the "development ethos" and the contemporary industrial mission.

Gwyn, Richard. "You, the taxpayer, pick up the tab for oil exploration." *Toronto Star*, August 30, 1979.
One of many articles on the financing of arctic development, arising out of a "textbook example of investigative journalism" by the *Calgary Herald*. An example of the continuing service to the public interest by the business and political press.

Hockett, Charles F., and Robert Ascher. "The Human Revolution." *Current Anthropology* 5(1964): 135–168.
The original statement and explication of "Romer's Rule".

Kapila, Sunita. "Underdevelopment and Development." Unpublished paper, Faculty of Environmental Studies, York University, 1979.
   Kapila, a Kenyan graduate student of Asian descent, has greatly helped the author's understanding of the process of cultural assimilation by way of industrial development.

Kelsall, John P. *The Migratory Barren-Ground Caribou of Canada.* Ottawa: Department of Indian Affairs and Northern Development, 1968.
   The major work. There are numerous subsequent papers and articles, some published by the Canadian Wildlife Service, others as research reports for environmental assessment purposes.

Kormondy, Edward J. *Concepts of Ecology.* Englewood Cliffs: Prentice-Hall, 1969.
   One of the "Concepts of Modern Biology" series, a useful basic summary for the beginner or for those who would like one manageable, modest text at hand. Like all such texts, not strong on northern ecology.

Lang, Reg, and Audrey Armour. *Environmental Planning Resourcebook.* Montreal: Environmental Canada (Lands Directorate), 1980.
   A comprehensive and meticulously organized collection of sources, examples, case histories and perspectives from the Canadian experience in environmental planning, resources, and management. A quite extraordinary assembly of useful data and pointers. Not directly applicable to arctic affairs, but much of the material would be transferable with little adjustment.

Leiss, William. *The Domination of Nature.* New York: Braziller, 1972.
   A seminal, critical study of the notion of the "conquest of nature" through science and technology, with emphasis on its seventeenth century origins.

_____. *The Limits to Satisfaction.* Toronto: University of Toronto Press, 1976.
   This essay on "the problem of needs and commodities" in contemporary society is of special interest in the matter of the production-consumption imperative which propels the industrialization of the arctic.

_____, ed. *Ecology Versus Politics in Canada.* Toronto: University of Toronto Press, 1979.
   An enlightening collection of original and provocative essays on the politics of Canadian environmental issues, from the points of view of biology, political science, economics, urban and regional planning, law, philosophy, and administration.

Livingston, John A. *Canada.* Toronto: Natural Science of Canada, 1970.
   The summary (ninth) volume in the Illustrated Natural History of Canada series, in which the arctic is viewed in the larger continental and circumpolar contexts.

_____. *One Cosmic Instant.* Toronto: McClelland and Stewart, 1973.
   Subtitled "A natural history of human arrogance"; a review of the origins and evolution of the anthropocentrism of western cultures, with some prehistoric speculations.

151

———. "Birds." Background paper on birds of the proposed pipeline route presented to Overview Hearings, Mackenzie Valley Pipeline Inquiry, March 1975.

———. "Communicating Wildlife Values", in *Proceedings of the Federal-Provincial Wildlife Conference 1975*. Ottawa: Canadian Wildlife Service, 1975.
Values and the valuing process in the face of the traditional utilitarian imperative.

———. "On the relevance of lungfish, lilacs, wolves, and spirit levels in resource-constrained economies."
Journal of Soil and Water Conservation, vol. 35, no. 4 (July-August, 1980).

———. *The Fallacy of Wildlife Conservation*. Toronto: McClelland and Stewart, 1981.
A study of the self-defeating nature of conventional arguments for wildlife conservation, with attention to root conceptual problems. Alternative perceptions and experience of the man-nature relationship are proposed.

Lopez, Barry Holstun. *Of Wolves and Men*. New York: Charles Scribner's Sons, 1978.
Poetically expressed, a deeply moving account of the ancient relationships between wolves themselves, between wolves and their prey and other aspects of their environments, and between wolves and both indigenous human cultures and our own.

Milne, Allen R., and Brian D. Smiley. *Offshore Drilling in Lancaster Sound: Possible Environmental Hazards*. Sidney, B.C.: Department of Fisheries and Environment, 1978.
Probably the most valuable source available. Detailed summaries of present knowledge of offshore drilling procedures, hazards, and constraints to drilling; fate of crude oil from a sea-bottom blowout; countermeasures to oil from a blowout; marine plants and animals of Lancaster Sound. Indispensable background in these and other matters. Especially valuable for its exhaustive bibliographical work.

Nettleship, David N., Timothy R. Birkhead, and Anthony J. Gaston. "Reproductive failure among arctic seabirds associated with unusual ice conditions in Lancaster Sound 1978." Dartmouth: Canadian Wildlife Service, 1979.
The murre disaster caused by weather in 1978 was a prime example of arctic unpredictability, and of the stresses under which wildlife must live at the best of times.

Odum, Eugene P. *Fundamentals of Ecology*. Toronto: W.B. Saunders, 1971.
The most widely-used and professionally recognized text in the New Ecology, by its most distinguished contemporary proponent.

Ophuls, William. *Ecology and the Politics of Scarcity*. San Francisco: W.H. Freeman, 1977.
> The best account and analysis yet published of the inseparable relationship between diminishing industrial resources and diminishing political and social choices. Do we restrain our industries and ourselves voluntarily, or do we let either the forces of nature or of unavoidable totalitarianism do it for us? The choice may still remain, but the "steady-state" approaches. This splendid work is further enhanced by unusually full and much elaborated bibliographic notes.

Paddock, William and Paul. *Famine 1975!* Toronto: Little, Brown, 1967.
> Includes the principle of *triage*, in this case applied to world human overpopulation, famine, and Third World aid.

Paradise, Scott. "The Vandal Ideology." *The Nation*, December 29, 1969.
> A notable summary of the cultural beliefs that underlie our industrial system, with suggestions as to their modification.

Pimlott, Douglas H. "Offshore Drilling in the Canadian North: Elements of a Case History", in *Mackenzie Delta: Priorities and Alternatives*. Ottawa: CARC, 1976.
> A conference presentation of insights gained regarding the approach of government and industry to the development of petroleum resources in offshore areas of the Canadian arctic during the preparation of *Oil Under the Ice* (see below), by one who knew more of such matters than most. A thoroughly chilling story.

Pimlott, Douglas H., Kitson M. Vincent, and Christine E. McKnight. *Arctic Alternatives*. Ottawa, CARC, 1973.
> Proceedings of the first CARC-sponsored workshop, on "people, resources, and the environment north of 60°" Summary backgrounds on perspectives, options, northern peoples, resources, the biophysical environment, and legal aspects of arctic questions. Much new information has come to hand since this was published, but it remains a solid, basic starting-point.

Pimlott, Douglas H., Dougald Brown, and Kenneth P. Sam. *Oil Under the Ice*. Ottawa: CARC, 1976.
> Result of an exhaustive and intensive examination of the history, procedures and practices of offshore drilling, and its environmental implications. Meticulously documented, the work reveals not only the apparent hollowness of expressed government concerns for northern people and environments, but also disturbing evidence of government secrecy in northern affairs. Required.

Reed, John C., and John E. Slater, eds. *The Coast and Shelf of the Beaufort Sea*. Arlington, Virginia: The Arctic Institute of North America, 1975.
> Proceedings of a 1974 conference designed to outline the then state of scientific knowledge of the region, and to define existing problems, toward a synthesis of Beaufort Sea environments and processes. Although more

work has been done in the intervening years, this is a useful background, especially for physical factors. Technical editor Wade W. Gunn.

Roszak, Theodore. *Where the Wasteland Ends*. Garden City: Doubleday, 1972.
A profoundly thoughtful and deeply spiritual examination of the interplay of politics, culture, and conventional perceptions of reality. Especially relevant to our purposes here is the discussion of contemporary science and technology.

Schenkel, Clara. "Identity." *Northern Perspectives*, vol. 7, no. 4, 1979.
A succinct editorial on the cultural melting-pot and its meaning for native peoples.

Schindler, David W. "The Impact Statement Boondoggle." *Science* 192 (1976): 4239
The oft-quoted condemnation of much environmental impact assessment by a biologist concerned for the well-being of both science and environments.

Schorger, A.W. *The Passenger Pigeon*. Madison: University of Wisconsin Press, 1955.
This account of the pigeon's natural history and extinction is of special interest in the study of incremental reduction of populations toward the eventual collapse of an entire species (that once numbered in the thousands of millions).

Schumacher, E.F. *Small is Beautiful*. New York: Harper and Row, 1973.
The contemporary classic by a maverick economist who added both a phrase to our language and a concept to our culture, and who called attention to the absurdity and presumptiousness of long-term forecasting, and to the necessity of a "becoming" existence through the development of "intermediate technology". This work is a touchstone.

Shepard, Paul, and Daniel McKinley, eds. *The Subversive Science*. Boston: Houghton Mifflin, 1969.
This collection of famous essays "toward an ecology of man" belongs in the library of all who would understand more than the external symptoms of environmental problems. The "subversive science" is ecology; the essays are about the ecology of man as a biologic, philosophic and technologic being, and about his relationships with both his natural and his manufactured habitats.

_____. *ENVIRON/MENTAL Essays on the Planet as a Home*. Boston: Houghton Mifflin, 1971.
A further collection, with central emphasis less on his ecology than on man himself, with attention to social and psychological, emotional and aesthetic questions. This and its predecessor are the best anthologies around.

Selected References

Singer, Peter. *Animal Liberation*. New York: Avon Books, 1977.
Drawing on examples from factory farming, the research laboratory, and other human activities involving animal suffering on a massive scale, the author (by profession a philosophy scholar) explores the cultural and philosophical underpinnings of our attitudes and behaviour, and offers in the place of "speciesism" a new ethics for our treatment of non-human beings. This book is valuable in our reflections on "externalities".

Snyder, L.L. *Arctic Birds of Canada*. Toronto: University of Toronto Press, 1957.
Although much has been added to arctic ornithology—especially with regard to distributions—since this book's publication, it remains the only work devoted exclusively to birds of the Canadian arctic. Detailed, readable text and useful maps. Stunning line illustrations by T.M. Shortt.

Stonehouse, Bernard. *Animals of the Arctic*. London: Ward Lock, 1971.
Subtitled "the ecology of the Far North" (a claim which is just a trifle sweeping), this generously illustrated book will be unusual to Canadian readers for its circumpolar (European-Asian-American) approach, which is refreshing in itself, and thus includes a number of species we do not encounter in Canada. A good, though abbreviated, introduction.

Symington, Fraser. *Tuktu: the Caribou of the Northern Mainland*. Ottawa: Department of Northern Affairs and National Resources, 1965.
Much research on caribou has been published since 1965, but this attractively written contribution remains of interest for its pleasant readability and its historic and social references.

Tener, J.S. *Muskoxen in Canada: a biological and taxonomic review*. Ottawa: Department of Northern Affairs and National Resources, 1965.
The recognized source. There are subsequent unpublished studies from individual arctic islands.

Torgerson, Douglas. *Industrialization and Assessment*. Toronto: York University Publications in Northern Studies, 1980.
Social impact assessment as a social phenomenon. Examines the social context as well as the theoretical content of SIA, with particular emphasis on the Berger Inquiry. Includes an important discussion of the "ideology of industrialization".

Tuck, Leslie M. *The Murres*. Ottawa: Department of Northern Affairs and National Resources, 1960.
The ranking work. For other references see Milne and Smiley (1978).

White, Lynn Jr. "The Historical Roots of our Ecologic Crisis." *Science* 155 (1968): 1203–1207.
This much-reprinted cornerstone contribution to modern ecophilosophy examines the role of the Judaeo-Christian tradition of anthropocentricity and its consequences.

155

Wilkinson, Douglas. *The Arctic Coast*. Toronto: Natural Science of Canada, 1970.
Another of the Illustrated Natural History of Canada series. A concise yet detailed popular account of physical and biological phenomena north of the limit of trees, with particularly interesting cultural descriptions. Very handsomely illustrated.

Woodford, James. *The Violated Vision: the rape of Canada's north*. Toronto: McClelland and Stewart, 1972.
A passionate account of the early years of the industrial invasion of the north, with emphasis on the "politics" of government-corporate relationships, and the environmental insensitivity of the development ethic. The arctic is seen as the ultimate testing ground of the much-advertised ethic of the "age of ecology".

Worster, Donald. *Nature's Economy: the roots of ecology*. Garden City, New York: Anchor Books, 1979.
A superbly researched and eloquently expressed account of the changing socio-cultural crucible out of which modern science—in this case ecology—has come. The contemporary New Ecology, with its economic bias, is traced to its origins in pre-Darwinian times by a historian of ideas with rare understanding of the complex intellectual and philosophical underpinnings of modern biology. Fundamental to critical appraisal of, for example, environmental impact assessment. For those who are interested in the fact that modern ecology is but one of many optional ways of understanding nature and natural process, required.

INDEX

157